BIG BAT YEAR

BIG BAT YEAR

A Conservation Story

NILS BOUILLARD

PELAGIC PUBLISHING

First published in 2023 by
Pelagic Publishing
20–22 Wenlock Road
London N1 7GU, UK

www.pelagicpublishing.com

Big Bat Year: A Conservation Story

A CIP record for this book is available from the British Library

https://doi.org/10.53061/JVJH4376

ISBN 978-1-78427-310-1 *Paperback*
ISBN 978-1-78427-311-8 *ePub*
ISBN 978-1-78427-312-5 *PDF*

Cover image: Australian Ghost Bat *Macroderma gigas* © Xavi René

Printed in the Czech Republic by Finidr

Contents

Foreword by Merlin Tuttle viii

Tenerife, Canary Islands, May 2018 1

New Zealand, December 2018 7

Pureora, North Island, New Zealand, January 2019 10

Invasive species 13

Viti Levu, Fiji, January 2019 17

Guadalcanal, Solomon Islands, January 2019 23

Island bats 28

Grande Terre, New Caledonia, January 2019 30

Queensland, Australia, January 2019 34

Darwin, Australia, February 2019 43

Echolocation 46

Subic Bay, Luzon, Philippines, Valentine's Day 2019 49

Short interlude: Diving break, Sulawesi, Indonesia, March 2019 54

Tangkoko, Indonesia, March 2019 57

West Papua, Indonesia, March 2019 63

How to become Bat(wo)man – aka a bat conservationist, aka a batter 69

Sepilok, Borneo, April 2019 71

Evolution 75

Taman Negara, Malaysia, April 2019 78
Kaeng Krachan, Thailand, April 2019 81
Reproduction 86
Bengaluru, India, May 2019 88
Mumbai, India, May 2019 91
Chengdu, China, May 2019 95
Bats and disease 98
Taiwan, June 2019 101
Okinawa, Japan, June 2019 108
Bats and culture 110
Virelles, Belgium, July 2014 112
Austin, USA, June 2019 116
Wind turbines 121
Sinaloa, Mexico, June 2019 124
Jalisco, Mexico, June 2019 126
Yucatán, Mexico, July 2019 129
Vampires and people 137
Antsiranana, Madagascar, July 2019 139
Andasibe, Madagascar, July 2019 142
Nairobi, Kenya, August 2019 146
St Lucia, South Africa, August 2019 152
Taxonomy 158
Mahé, Seychelles, November 2018 160
Tel Aviv, Israel, September 2019 163
Eilat, Israel, September 2019 165
Bat migration 169
Puerto Maldonado, Peru, November 2019 171

Food habits 178

Galápagos, Ecuador, December 2019 181

Rio Claro, Costa Rica, December 2019 192

Bats and ecotourism 202

Brussels, Belgium, January 2020 204

Acknowledgements 207

Index 209

Foreword

by Merlin Tuttle

Nils Bouillard has a passion for nature and adventure. Through his round-the-world search for seldom-seen bats and birds, he shares countless discoveries, from brilliantly colorful pit viper snakes to flying lemurs, even seadragons and millipedes. His scariest adventure, being attacked by a vicious dog, occurred while riding on the back of a motor scooter in India. For the average reader, he'll turn the world upside down – he reports being terrified of dogs but loves bats!

While just a 23-year-old working on a Master's degree, he decides to spend a year traveling the globe in search of bats. To his surprise, his parents not only don't think him crazy, but they also wisely agree to support him in perhaps the most valuable education obtainable: firsthand experience. He learns to cope with a wide variety of challenges – torrential rains, sweltering deserts, communication difficulties, unfamiliar customs, and travel delays.

Nils' original goal was to see 400–500 species of bat. However, he quickly realized that his goal was more than a little ambitious. One of the challenges lay simply in finding bat-watching guides. Also, bats could be difficult to find, given their mostly nocturnal lifestyles. Based on early experience, he lowered his expectations, hoping to reach 365 species, an average of one per day. It didn't take long to discover that, despite the scarcity of bat-watching guides, birders often found bats equally fascinating, and were eager to learn.

Given his intense curiosity about virtually all living things, Nils was seldom bored. In Fiji, he was delayed several days by a cyclone. Its torrential rains and 160+ km/h winds washed out and blocked roads. But slow going had its rewards. He was thrilled to see magnificent Golden Doves, Fiji Bush-warblers, and small endemic birds referred to as Slaty Monarchs. The grand prize was a Samoan Flying Fox, endemic to Samoa and Fiji. This handsome, three-foot-wingspan creature is one of the world's most diurnal bat species, an important pollinator and seed disperser.

Nils soon realized that, even to acquire a list of 365 identified species, he would also have to include bats identified only by their calls. Yes – as he explains, many bats can be identified by their unique ultrasonic vocalizations. Nevertheless, many haven't yet been recorded, and others haven't even been discovered. Some are too similar to be identified or rely on signals so faint that they are called "whispering bats." Fortunately, at most locations, at least a few species used calls so loud and unique that they could be identified. He would add species to his list only when he personally saw or recorded them in the wild, or on lucky occasions, when he could accompany local researchers capturing bats for study. Great care was taken to minimize disturbance.

Nils first became fascinated by bats when he met several close up during a field trip for young people, sponsored by the Natagora-Jeunes youth club in Virelles, Belgium. He was immediately intrigued by the amazing diversity and worldwide importance of bats. But unlike most people, he was especially curious to learn about their sophisticated use of echolocation to navigate. Even better, they could often be identified by their unique calls. Nils' around-the-world bat junket began due to a combination of extreme fascination with these creatures and his love of extra-big challenges.

Readers wishing to learn more about these elusive animals will enjoy the special effort Nils has taken to explain interesting adaptations and behaviors, in sidebar boxes strategically located between stories of adventure. The sidebar information is carefully researched and clearly explained, covering diverse topics from reproduction, feeding and migration to threats, disease and taxonomy.

You'll share in his excitement at meeting flying foxes as large as eagles and Bumblebee Bats that weigh less than 2 g (the same as a U.S. penny), and at discovering stunningly bright-colored painted bats in Asia and Yellow-Winged Bats in Africa. His finds include giant Naked Bats in Malaysia, the world's largest insect-eaters, and Horseshoe Bats as strange as any dinosaur, in the Philippines. In his year-long quest you will meet some of the world's rarest, strangest, and most important species, and hopefully will want to learn more about these elusive, long-misunderstood mammals.

This isn't just the story of an incredibly challenging trip, limited to reporting of the 396 bat species he found. It's also about visits to out-of-the-way places you've never heard of before, and the unusual foods, customs and other wildlife Nils excitedly observes and describes. His bird list is impressive. And he even reports the thrill of finding his first seadragon, a tiny seahorse, while scuba-diving off the coast of Indonesia.

This book provides a fun record of overcoming setbacks, finding unexpected bonus opportunities and meeting one's goals. Nils admits to the disappointment of seeing vast deforestation and the destruction of unique island faunas, and mentions the sickening carnage he encountered in Indonesian markets where some 500 tons of fruit bats are imported annually, cramped in deplorable misery in tiny cages, then sold as food delicacies. He shares his disgust but doesn't dwell on it. Overall, this is a highly informative tale of adventure and discovery, of broad interest to nature enthusiasts everywhere.

January 2022

Tenerife, Canary Islands, May 2018

There I was, lying in the bottom bunk of my bed at Drago Nest Hostel on the outskirts of Puerto de la Cruz, Tenerife. Little did I know at the time that the decision I'd make there and then would change many things in my life. It's often hard to pinpoint specific life-changing moments – in a way, all our decisions are life-changing, and we never get to see the alternative outcomes. But I can't really think of many other specific junctures in my life that have radically shaped the course of the rest of it. Perhaps this shouldn't surprise anyone, considering I was only 24.

It started with the seemingly harmless idea of going on holiday, to take a break from my stressful Master's at Imperial College London. The Canary Islands had been on my wishlist for a long time and I decided to take the opportunity to tick that one off.

Tenerife, a hostel, a bunk bed – let's get back to it, or rather the few hours preceding that exact moment. I had been going through some mammal-watching reports for hours, trying to figure out if doing a Big Bat Year was feasible. Well, to be exact, I knew it was, but I didn't know whether it would be exciting. Big Years can be done anywhere and with any group of animals, but not all appeal to a wide audience. And without finding a significant number of species, it might be a rather dull enterprise. The more I read, however, the more I wanted to do it. I was becoming increasingly convinced that a Big Bat Year could be a worthy pursuit – that people who wouldn't normally care about bats, for example birders, could become interested in them with me because of the familiar nature of the endeavour.

On my first morning on Fuerteventura, I drove straight to an area of arid habitat known to harbour most of the island endemics. I successfully found a couple of Cream-coloured Coursers, a strange wader that strays away from the mudflats that most of its cousins love and settles for the far less hospitable deserts. It was my first tick on the island – 'tick' being a word used to refer to a new addition to one's species list. Needless to say, it's something I would search for very often during my Big Bat Year.

After a reasonably long drive around the island, I'd seen most of the species I'd come for, except the Bustard. I'd had some fun time with some Barbary Ground Squirrels that, despite the signs forbidding anyone to give them food, were obviously expecting me to give them something in return for their poses for the camera. The island doesn't have many native mammals, apart from bats (as is the case with many islands around the world, actually), but some exotic-looking creatures have been introduced there and are thriving. The squirrel is one; the hedgehog is another. Unfortunately, I did not see live hedgehogs.

I used the high heat of the afternoon as an excuse to go back to my hostel, and back to my Big Bat Year planning. The name of my journey came to me quite naturally. I was about to start a Big Year, and its main distinctive feature was, of course, that it would focus on bats. . After many hours of picturing myself in remote parts of the world, looking for rare bat species from behind my laptop, I headed back out. Before it was completely dark, I headed back to the place I had visited in the morning, hoping the Bustards would show themselves – and they did! One even crossed the road. I'd also spotted an area that had some rocky outcrops that I was hoping could be home to some bats, and so I spent most of my evening there.

My recorder was showing me exciting signs of ultrasonic activity. Could this be the European Free-tailed Bat *Tadarida teniotis*, a species not known to occur on this island but which had been found on neighbouring Tenerife? Well… no. The source of the sound was none other than my cute little Hyundai i10 making some funny ultrasonic noises. It turns out cars have an entire repertoire of annoying sounds that one can pick up with a bat recorder. I hadn't seen a rare species after all, but my Big Bat Year hadn't started yet.

The next day, my second and last day on the island, I didn't feel the need to wake up early, as I had seen most of the birds I wanted to see. If I had to highlight one thing I didn't miss when moving from birds to bats, it would be the early mornings. I can't do those. One could call me lazy, but I prefer to think of myself as an honorary bat – although not being able to sleep through the winter is a bit of a let-down.

Naturally, that day I spent my time indoors looking up some more reports. I'd downloaded most of what the internet had to offer on bats of Oceania and Asia, and I was now looking into African bats. The back-of-the-envelope maths I'd started doing to calculate how many species I thought I could score was edging close to a number I thought was worth it. I'd estimated I could tick 400–500 species, based on the reports I'd already collected and flicked through.

What better place to plan the Big Bat Year than amidst these stunning Canary Island landscapes?

There was one more thing I wanted to do on this island, and that was to investigate a previous report of the Tenerife Long-eared Bat *Plecotus teneriffae*, a species restricted to the western Canary Islands. Once I was in front of the church where it had apparently been sighted, looking into every crevice, I came to an all too important realisation: reports aren't always accurate and even when they are, they don't guarantee another sighting. I stopped for some dragonflies on the way back. They, unlike bats, rarely disappoint, but they can be awfully predictable.

That day made me realise that perhaps aiming for 400–500 bat species was a bit ambitious. Maybe aiming for one new species a day was more reasonable – so a total of 365 bat species in as many days. When I got back to my hostel, I spent some more time on my laptop, drawing up a list of countries I wanted to visit and outlining a rough itinerary. When that was done, I'd taken my decision. I wanted to do this.

I called my parents later that day and told them about my crazy idea. They didn't think it was crazy, or maybe it was just that they're used to my level of crazy. I had laid the first stone in the foundation of my adventure. The next steps were to plan the trip and figure out the rules I'd follow for this 'competition'. As I was the first individual ever to attempt it, I could set my own rules, but I didn't want to make things too easy for myself either. After all, this was to be my next big challenge. The rules I felt comfortable holding myself to were the following:

- Similarly to birders counting 'heard-only' birds, I decided to count sound-recorded bats. Hearing is as valid a sense as sight, so there's no point dismissing it. Many bats can only be identified either in the hand or from their echolocation. Additionally, shining a torch on a bat to record it didn't seem justified. I was far more likely to disturb the bat than I was to get any useful information out of it.
- As an extension to the previous rule, I decided not to count visual sightings of bats in flight (unless it was something obvious such as a flying fox). Most of the time, a second or two of a bat flying across a trail is only enough to narrow it down to a couple of species. Trip reports are filled with tentative IDs of bats in flight; most of them are unverifiable.
- I had to be present for a bat to be counted, meaning that no static recorders could be used. This is an obvious one – but given I'd consider sound recordings as valid, I had to draw the line somewhere. I did, however, still use static recorders in some places. I wouldn't count the species I got using this method for my Big Bat Year purposes, but the collected data was still valuable for local projects.
- Under no circumstances was I to send away all my recorders or photos to someone for them to ID the species. Getting assistance was acceptable but I had to take the lead on all identifications.
- The bats had to be wild – so no bats in zoos, obviously, but also no bats in rehabilitation, not even upon release.
- There have been no invasive or non-native bat populations (except perhaps for the Japanese Pipistrelle *Pipistrellus abramus*, on Okinawa – depending on the sources you check) since the extinction of the Egyptian Fruit Bat *Rousettus aegyptiacus*, on Tenerife. Therefore, I didn't have to worry about having a rule against counting anything non-native.

The following weeks were spent chiefly on additional planning: looking up reports, finding local contacts and guides, as well as collecting as many scientific papers as I could. Unlike bird tourism, bat tourism isn't really a thing, so finding accurate travel reports is often impossible. The vast majority of location data for bats is only available in articles behind paywalls. Luckily, I was still a student at one of the world's top universities, so I had access to most of them.

I say I was still a student, but I didn't spend the majority of my time on my degree research project, to be perfectly honest. I was experiencing some rather annoying delays in getting my data from a centralised European database (my lab ended up receiving the data three months after I graduated), so it's fair to say I was struggling somewhat. I always have a grievous relationship with anything I consider a failure – that is, anything that I'm responsible for that doesn't work as I want it to. This led to some valuable discussion with my parents about whether the Big Bat Year was a good idea. I was bound to hit some walls, lots of them in fact. However, all the time spent planning my journey gave me a way to escape the stress of my degree. It genuinely helped me cope a lot better than I could before.

As I slowly reached what seemed like the end of the available resources I could use to plan my journey, I started thinking about the branding and my communication channels. I'd never had a website before, or a logo, or a brand of any sort. It was all new to me, but I knew I wanted it to be eye-catching. My logo ended up taking me several dozen hours of work, primarily spent in tweaking the colours used or the width of some lines. But if I was to procrastinate, I might as well do it right! Finding the right colours is no easy task – I'm sure any graphic designer would agree. I love blue, so I knew that's what I was going for, but finding a light, medium and dark blue that fit together without it looking like I was opening a spa wasn't easy. To this day, I am still convinced that spending five-plus hours on finding the right colours was worthwhile. Perhaps that would explain why I tend to go back to those colours whenever I need to design some graphics.

A few days before I was due to submit my thesis, I found nothing better to do than heading to the infamous British Birdfair. The British Birdfair is the annual meeting of the birding community – thousands of people visiting from dozens of countries, all sharing the same passion. It was my second time attending this event, and unlike my first time, I now had a purpose: finding people to help me on my quest. Lodge owners and tour guides were my last possible sources of information to plan this journey as well as I could. There were only so many trip reports I could go through to collect data, and while scientific papers were a great source of information too, they can be a pain to get access to and rarely make for exciting reading because of the jargon. So this turned out to be a nice change, being able to chat with others about where to find bats.

It was also the first time I started getting a feel for whether people would be excited about my idea. It turned out that even birders were intrigued, and it was probably at that point that the true reasons behind the trip became apparent. I wasn't doing this because a friend of mine had jokingly said it would be a good idea; I was doing it because it excited both myself and the people around me. At that point in my life, I'd never had that feeling before – the sense of belonging to a group of like-minded people encouraging the wildest ideas. Knowing I'd have dozens, if not hundreds, of people following my travels and keeping me going gave me the confirmation I needed. I got in touch with Round the World Flights, a British travel agency specialising in multi-stop tickets, a few days later. Given the sheer number of flights I'd be boarding, I was willing to pay for some assistance, in case anything should go wrong when I was in the jungle. It turned out to be a brilliant move – my contact Stuart could always sort out any issues arising in record time. I can't thank him enough.

Speaking of flights, because I had decided to book all the international flights well ahead of time, mainly to save on cost, a large portion of my itinerary would be set in stone. This was an additional challenge I was giving myself, but it came with the peace of mind of not having to deal with the tickets or the airlines myself. That trade-off was one I was happy to live with.

How I managed to juggle my research project and the Big Bat Year planning simultaneously, I'm not quite sure. All I know is that, had I not also been working on the Big Bat Year to give myself something to look forward to, I may never have finished my Master's degree.

New Zealand, December 2018

While New Zealand is undoubtedly my favourite country in the world, it has very few bats – with just two species. That is why I had to compromise heavily on the time I could spend there. My solution to this very significant problem was to start my journey in New Zealand. That way, I could spend a good two weeks among its wonderful landscapes before my Big Bat Year officially started, after which I'd only have an additional four days. When I landed, I felt like I was coming back home. I'd been there only once before, but it somehow all felt so familiar that I wasn't stressed at all. This was another reason I thought starting in New Zealand would be a great idea.

I spent a few days up north on my own before being joined by Manu, an old friend of mine who decided to accompany me as often as he could during my world tour. Together, we spent a lot of time birding; I showed him all the local specialities, including the urban penguins in Oamaru. Oamaru is a small coastal town on the east coast of the South Island that's famous for its penguins. Some have reportedly even visited the local pub (sadly, we didn't see any engaged in this activity). The main focus of our trip to the South Island, however, was the Eglinton Valley in Fiordland, on the west coast.

Eglinton Valley has been the focus of many conservation projects as its almost pristine beech forests are home to some of the rarest species in the country, both birds and bats. These New Zealand beech forests (not to be confused with the Northern Hemisphere beech trees; they're quite different) are the remnants of the ancient cold and wet forests of the Gondwana era (about 180 million years ago) and have a distinctive fairy-tale feeling. It's easy to see why Peter Jackson picked this kind of forest to represent Elvish settlements such as Lothlórien in *The Lord of the Rings*. Even without CGI, you'd expect elves to come out and greet you (if you're friendly, that is). However, these forests now have a dark side; they're plagued with invasive species such as possums, rats and wasps that not only destabilise the ecosystem but actively and rapidly contribute to the extinction of countless native species. Only through drastic conservation measures such as culling have these forests retained some of the most unique species in New Zealand.

Most people visiting Fiordland only drive through this valley on their way to a cruise on the Milford Sound. The valley itself is probably one of my favourite places in New Zealand; the scenery is exceptional and the wildlife equally so. This valley is still home to many native species doing poorly elsewhere, such as the Yellowhead/Mohua, the Southern Brown Kiwi/Tokoeka, the Blue Duck/Whio and Pekapeka, the native bats. Like the forests I'd visited the week before on the North Island (though I didn't find any bats there), this valley is home to both bat species, so it was a must-visit location. Those forests are magical, filled with moss and ferns everywhere you look. Every level of the forest has its range of fern species, and anything that isn't covered by some odd-looking example such as the Kidney Fern is covered in moss.

We left in the middle of the afternoon to ensure we'd have time for some birding before 'bat-time'. Our targets were mainly the Rock Wren, which I hadn't seen yet, and also Yellowhead, Kaka, Pipipi (also known locally as Brown Creeper) and the South Island Robin, which Manu hadn't seen. We managed to score the last three. There was no Yellowhead to be found, but I'd never actually seen it there myself. I had only seen it on sanctuary islands such as Ulva and Blumine/Oruawairua, where thanks to successful reintroductions it is much easier to spot.

After an unsuccessful bat search at dusk at Lake Gunn, we headed to Mirror Lakes. Right before we arrived, a bat flew in front of the car! I jumped out and grabbed my recorder to find out what species it was. As mentioned above, I don't believe in bat identifications from flight alone, especially not when one dashes in front of the car like that. We didn't have to wait very long before the bat came back and showed up on the screen of my recorder, proving that it was a New Zealand Long-tailed Bat, the more common of the area's two remaining species. It wasn't a Short-tailed Bat, which was the one I really wanted to see – but still, I was over the moon. This was the first bat of the Big Bat Year. But also not really, because we weren't in 2019 yet... Well, it was a warm-up.

The Long-tailed Bat *Chalinolobus tuberculatus* belongs to a genus widespread across Oceania and quite similar to the *Pipistrellus* I was familiar with. The Short-tailed Bat *Mystacina tuberculata*, on the other hand, is very distinctive. This species is the last extant representative of the genus since *Mystacina robusta* became extinct in the 1960s. It's well adapted to life on the ground, with powerful hind legs, and is known to forage mainly by running around the forest floor, looking for flowers as well as insects. It's easy to understand why this was the species I most wanted to see. But there we were, close to roosts of both *Mystacina* and *Chalinolobus*, and only the latter showed itself that evening, unfortunately. I knew I'd have another opportunity to see both species on the North Island, and those would count for the Big Bat Year list.

Because of scheduling issues, I wasn't able to position myself in a bat habitat at midnight on 31 December. Instead, as 2019 began, we had to wait another couple of days and travel north to Pureora Forest Park, another restored woodland area in the centre of the North Island.

Pureora, North Island, New Zealand, January 2019

The Big Bat Year had begun!

I'd arranged to meet up with a team from the Department of Conservation (DOC) by getting in touch with Abi Quinnell, who had planned a trapping survey in Pureora targeting both species of bats. It was a fantastic opportunity for me to get up close and personal with the bats without causing any additional disturbance, as this was a regular survey by DOC. I was very careful not to disturb the bats, avoiding the roosts I knew of, as the bats are already under much stress due to the presence of introduced predators such as rats and stoats. I did not have the opportunity to visit Whenua-Hou, where the Short-tailed Bats appear to be thriving. Realistically, this would have made my quest far easier, but I find DOC's policy of limiting visits to those islands quite reasonable. The team had placed a harp trap across a small stream earlier that afternoon. Harp traps consist of rows of fishing lines stretched between two horizontal bars. They form a sort of screen through which bats try to fly because they're unable to accurately detect the barrier formed by the fishing lines. Most bats fail to pass through the second bank, but some particularly agile ones have been shown to fly through three or even four banks and escape on the other side. Those that don't manage to escape slide down the fishing lines and into a large canvas bag at the bottom, where they can calmly await rescue by researchers.

One of the main benefits of harp traps compared to mist-nets is that they can be deployed ahead of time with minimal risk of catching birds. They also appear to be more effective at catching certain species, and they make it a lot easier to extract the bats as they simply sit in a large fabric pouch at the bottom. It's all nowhere near as stressful as a mist-net, for both the bats and the surveyors. I'd used a harp trap before when I went on a trapping survey in the UK, but I didn't have much experience because we don't use them in Belgium. It is true that they

Catching bats with the Department of Conservation in New Zealand.

are expensive and bulky but they offer useful benefits to bat workers and bats alike. I'm not sure why they're not more prevalent in continental Europe compared to the UK, where many bat groups has one (or several). While we waited for the sun to set, Abi told me she had something to show me that would probably make me smile.

I was curious, especially as I hadn't been out in the field with people doing conservation work in New Zealand before. She took a box out from under a table and handed me the cutest bat detector I'd ever laid eyes on. It had two buttons, for a low and a high frequency. No dials, no screen, certainly no bells and whistles. Perfectly suited to bat work in New Zealand, it was the Microbat bat detector made by Stag Electronics (now Batbox Ltd). I later learned that they had been in use in the UK for quite a while, too – primarily for bat walks and such, not so much for research. But since New Zealand only had two species, there was no point in spending thousands on more sophisticated recording devices.

To find bats that evening, though, I was still going to rely on my EchoMeter, so I put the detector back in the box, and we headed out for the forest trail where the trap had been set. Once we got there, we started unpacking our kit to set up the processing station, only to realise the bat bags had stayed in the truck. Abi decided to climb back to get them before the first bats were in. While we waited, we were told about the presence of Huhu Beetles *Prionoplus reticularis*, an endemic species of longhorn, attracted to light. They advised us that it was best to keep them at bay because their bites could be pretty painful, and they appeared to bite without provocation. Here I was thinking New Zealand had no nasty insects.

Not too long after that, we removed two Long-tailed Bats from the harp traps. I'd already seen and recorded this species on the South Island, but this one actually counted and officially became the first species on the Big Bat Year list. Exciting start! Well, apart from the Huhu beetles, which were genuinely annoying. After taking a couple of photos of *Chalinolobus*, we set them free and waited for more. Those didn't come, so Manu and I decided to head back on our own, hoping to record something else along the way. We also had a lengthy drive ahead of us, which is why we didn't want to stay too late.

Leaving the recorder on for the walk back to the car appeared unsuccessful, but we were focusing more on the hundreds of glow-worms than the recorder. The glow-worms put on a truly magical show. They are a famous attraction in New Zealand and tourists often spend quite a lot of money on cruises inside caves in Hamilton or Te Anau, because very few of them know that glow-worms can also be found in the forest, for free. Admittedly, they gather in greater numbers within the caves – but they can be quite common along trails, particularly on undercut banks when the overhang creates a darker shelter for them. Luckily, the recorder was paying attention and did pick up the calls of the quiet Short-tailed Bat *Mystacina tuberculata*. Unfortunately, I noticed it too late and was unable to spot this bizarre species that spends far more time on the ground than any other bat in the world.

Invasive species

Invasive species represent one of the top five threats to biodiversity, and vertebrates in particular, on Earth.[1] They are estimated to be responsible for as much as 40% of mammalian extinctions in recent times,[2] with most of these extinctions occurring on islands.[3] Islands are particularly vulnerable as non-native predators can quickly colonise the entire area, but the danger isn't restricted to them. The most widely discussed examples are those involving non-native predators, most notably cats and rats, but there are examples of competition between native and invasive species as well. This is most common among plant species but there are examples involving animals too, including bats.[4] Direct competition is described as two species fighting for the same resources. Indirect competition, on the other hand, is more subtle – for example, a plant species outcompeting another one when that other species is the primary food source for an animal. There are also records of invasive species outcompeting other native 'prey' species but not constituting a suitable replacement – such as the case of poisonous Cane Toads colonising numerous islands in the Pacific.[5]

Here are a few examples of the negative impact invasive species can have on bat populations around the world:

Raccoons in Europe

Raccoons were introduced to Europe in the 1920s through the fur trade. There are now many healthy, thriving populations of this North American species in Europe. It is a mainly arboreal predator that feeds on bird eggs, chicks, small mammals, amphibians and bats. It has been shown to visit underground sites containing hibernating bats in Poland and feed on the sleeping bats.[6] Bats are highly vulnerable to predation during hibernation, so the introduction of a well-adapted predator can have dramatic consequences.[7]

Rats and stoats in New Zealand

Rats have been introduced on many islands in the Pacific, including New Zealand, via boat. Three species are currently known to inhabit Aotearoa; the Polynesian Rat *Rattus exulans* was the first to be introduced, about 2,000 years ago. The other two species, the Ship Rat *Rattus rattus* and the Norwegian Rat *Rattus norvegicus*, were introduced much later. From the main islands, they then spread onto the smaller offshore ones – as was documented in 1962 when they colonised Big South Cape Island, decimating the last populations of Bush Wren *Xenicus longipes* and Greater Short-tailed Bats *Mystacina robusta*.[8] On the islands where they have been introduced, rats are opportunistic predators feeding on the eggs and chicks of seabirds, amphibians and reptiles, and on other mammals.[9] New Zealand bats, *Mystacina tuberculata* specifically, are endangered by introduced mammalian predators such as rats because of their partially terrestrial lifestyle.

Unlike rats, stoats were intentionally introduced to New Zealand to control the rabbit populations.[10] It turns out that they're equally effective at controlling native bird and bat populations. Predator eradication on offshore islands and diversity hotspots such as the Eglinton Valley has enabled the bat population's survival.[11]

Cane Toads and Ghost Bats

Cane Toads were introduced to Australia in 1953 as a means to control pest insects in sugar cane crops.[12] While certain species have learned to avoid Cane Toads or have even developed resistance to their toxin, this cannot be said of all.[13] Large-scale mortality attributed to a newly arrived toxic prey – that is, the Cane Toad – has been described in a variety of species across its non-native range. In some cases, the evidence is not definitive – but it's still considered likely that the Cane Toads are causing a decline. One such example is the Ghost Bat *Macroderma gigas*, Australia's largest carnivorous bat species. Like the rest of its family, it feeds on small mammals, including bats, reptiles and amphibians. This is a species that most likely attempts to feed on Cane Toads, and may be poisoned.[14]

Notes

1 Bellard, C., Genovesi, P. and Jeschke, J.M. (2016) Global patterns in threats to vertebrates by biological invasions. *Proceedings of the Royal Society B: Biological Sciences* 283 (1823): 20152454. https://doi.org/10.1098/rspb.2015.2454

2 Doherty, T.S., Glen, A.S., Nimmo, D.G., Ritchie, E.G. and Dickman, C.R. (2016) Invasive predators and global biodiversity loss. *Proceedings of the National Academy of Sciences* 113 (40): 11261–5. https://doi.org/10.1073/pnas.1602480113
3 Medina, F.M., Bonnaud, E., Vidal, E., Tershy, B.R., Zavaleta, E.S., Josh Donlan, C., Keitt, B.S., Le Corre, M., Horwath, S.V. and Nogales, M. (2011) A global review of the impacts of invasive cats on island endangered vertebrates. *Global Change Biology* 17 (11): 3503–10. https://doi.org/10.1111/j.1365-2486.2011.02464.x
4 Gurevitch, J. and Padilla, D.K. (2004) Are invasive species a major cause of extinctions?. *Trends in Ecology & Evolution* 19 (9): 470–4. https://doi.org/10.1016/j.tree.2004.07.005
5 Harvey, J.A., Ambavane, P., Williamson, M. and Diesmos, A. (2021) Evaluating the effects of the invasive cane toad (*Rhinella marina*) on island biodiversity, focusing on the Philippines. *Pacific Conservation Biology* 28 (3): 199–210. https://doi.org/10.1071/PC21012
6 Cichocki, J., Ważna, A., Bator-Kocoł, A., Lesiński, G., Grochowalska, R. and Bojarski, J. (2021) Predation of invasive raccoon (*Procyon lotor*) on hibernating bats in the Nietoperek reserve in Poland. *Mammalian Biology* 101 (1): 57–62. https://doi.org/10.1007/s42991-020-00087-x
7 Salgado, I. (2018) Is the raccoon (*Procyon lotor*) out of control in Europe?. *Biodiversity and Conservation* 27 (9): 2243–56. https://doi.org/10.1007/s10531-018-1535-9
8 Bell, E.A., Bell, B.D. and Merton, D.V. (2016) The legacy of Big South Cape: rat irruption to rat eradication. *New Zealand Journal of Ecology* 40 (2): 212–18. https://doi.org/10.20417/nzjecol.40.24
9 Jones, H.P., Tershy, B.R., Zavaleta, E.S., Croll, D.A., Keitt, B.S., Finkelstein, M.E. and Howald, G.R. (2008) Severity of the effects of invasive rats on seabirds: a global review. *Conservation Biology* 22 (1): 16–26. https://doi.org/10.1111/j.1523-1739.2007.00859.x
10 O'Donnell, C.F., Dilks, P.J. and Elliott, G.P. (1996) Control of a stoat (*Mustela erminea*) population irruption to enhance mohua (yellowhead) (*Mohoua ochrocephala*) breeding success in New Zealand. *New Zealand Journal of Zoology* 23 (3): 279–86. https://doi.org/10.1080/03014223.1996.9518086
11 Pryde, M.A., O'Donnell, C.F. and Barker, R.J. (2005) Factors influencing survival and long-term population viability of New Zealand long-tailed bats (*Chalinolobus tuberculatus*): implications for conservation. *Biological Conservation* 126 (2): 175–85. https://doi.org/10.1016/j.biocon.2005.05.006. O'Donnell, C.F., Edmonds, H. and Hoare, J.M. (2011) Survival of PIT-tagged lesser short-tailed bats (*Mystacina tuberculata*) through a pest control operation using the toxin pindone in bait stations. *New Zealand Journal of Ecology* 35 (3): 291–5.
12 Pikacha, P., Lavery, T. and Leung, L.K.P. (2015) What factors affect the density of cane toads (*Rhinella marina*) in the Solomon Islands? *Pacific Conservation Biology* 21 (3): 200–7. https://doi.org/10.1071/PC14918

13 Phillips, B.L. and Shine, R. (2006) An invasive species induces rapid adaptive change in a native predator: cane toads and black snakes in Australia. *Proceedings of the Royal Society B: Biological Sciences* 273 (1593): 1545–50. https://doi.org/10.1098/rspb.2006.3479

14 White, A.W., Morris, I., Madani, G. and Archer, M. (2016) Are cane toads *Rhinella marina* impacting ghost bats *Macroderma gigas* in northern Australia? *Australian Zoologist* 38 (2): 183–91. https://doi.org/10.7882/AZ.2016.028

Viti Levu, Fiji, January 2019

I'd kept a close eye on the weather forecast, because a cyclone was about to hit Viti Levu and could significantly alter my travelling plans; Manu had already gone back to Belgium by that point. Luckily, my concerns regarding the cyclone quickly vanished as the flight went by without a single moment of turbulence. After a bit of a struggle to locate one of the small buses that were replacing the large ones unable to drive due to the storms, I was on my way to Suva, the capital city. I wanted to be close enough to Suva that I could easily travel there to see the bats, but I was also hoping it would serve as a good base to easily explore the island using public transport.

We hadn't got very far before the first signs of the storm became apparent: fallen trees and dirt all over the road. About halfway to our destination, a large section of the road, almost half its width, had disappeared, taken by the torrential rains. The cyclone may have missed the island but its tail, and the accompanying tropical storms, hadn't. A little bit further down the road, we were forced to a halt. As we disembarked, we could see exactly why we'd had to stop: the front left wheel had come off. It wasn't just the tyre – the axle had broken. The locals tried their best to help me understand what was going on, but we all laughed when they realised one didn't have to be a car expert to know that something wasn't right. So, we waited for an hour or so for another van to pass and pick us up. What should have been a three-hour journey had turned into a five-hour drive by the time we reached the city. A short twenty-minute drive later, I was in the lobby of my accommodation, Colo-I-Suva Rainforest Eco Resort, in the hills above the city. The surroundings were incredible. The lodge is set in the middle of the forest, at the edge of a lake. They've placed the restaurant right on the shore of the lake, which meant that every meal I had was fantastic simply because I had lots of birds to look at. The lush and green surroundings were the reason I'd picked the lodge.

After settling in, I headed for the restaurant for my first taste of local food. I asked the waitress what she recommended I have on my first day in Fiji, and she said 'Kokoda' – fresh fish marinated in lime juice and coconut milk, served in a coconut. It was lovely. While I was lucky that the cyclone had missed Viti Levu, the island I was on, I wasn't so fortunate when it came to avoiding the torrential rain that comes with the tail of a cyclone. It was a small price to pay to avoid 160+ km/h winds, though! The following two days were therefore spent exclusively at the lodge, mostly between my room and the restaurant. Surprisingly, the birding was not bad, especially at the lake where I spent most of my time. In the trees on the lodge's grounds, I found a Giant Honeyeater, a bird endemic to Fiji that is known to be surprisingly easy to find. I also spent a lot of time playing with a couple of kittens and their mother, who lived on the grounds. The kittens were timid at first – unlike their mum, who was extremely affectionate. Spending two full days with the kittens helped them trust me a bit more, and soon they were happy to be near me as long as their mum was on my lap. Being Belgian, I thought I knew what rain was but those two days showed me I didn't. I messaged Ellie, my girlfriend at the time, saying it was 'Raining cats and dogs and elephants' – because that's what it felt like. Constant pouring rain.

On 8 January, the skies cleared, and I could finally get out somewhere. I only had two bat-watching locations to visit on the island. I decided to head to the easy one first, the Presidential Palace in Suva, down the road from where I was staying. I asked a staff member about the most straightforward way to get to the city, and she said it was either the bus or a taxi. Most guests would have chosen the cab, but I wanted to go on a local bus. So I waited a while for a bus to pass by and hopped on. I could see from people's faces that they weren't used to seeing white-skinned, blonde-haired and blue-eyed people all that often. I went to sit at the back, and a little kid stared at me for the whole journey.

Suva reminded me a lot of New Zealand, with all the banks and shops being the same. It was inside those shops that the differences became apparent. On the inside, they looked a lot more like Asian supermarkets than Western ones. This juxtaposition was fascinating. It didn't take me long to find trees containing bats, because those bats are relatively noisy. I was uncomfortable spending so much time with my camera and binoculars around the compound because I wasn't sure how friendly or understanding the guards would be. While Pacific Flying Foxes *Pteropus tonganus* were easy to find because they were numerous and noisy, I also wanted to see the more elusive Common Samoa Flying Fox *Pteropus samoensis*. It's a species more associated with forested habitats, but a few individuals are known to roost among the *P. tonganus*. They're of a very different colour, a combination of light brown and a slightly darker one, compared to the almost black fur of *tonganus* with their pronounced buffy collar. *Samoensis* were a lot harder to find, especially given how active the *tonganus* were, flying from one tree to another – it made checking each flying fox individually almost impossible. Eventually, I spotted a small gang of more discretely coloured flying foxes on the edge of the colony. The more I looked at them, the more I could see how they differed from the more common species, and I became convinced these were, in fact, *samoensis*. That was one more species for me, and the third species for the Big Bat Year list!

A typical Fijian landscape.

Getting worried that I'd spent close to an hour outside the compound but also satisfied with my excursion, I decided to move on and head for the seashore to spot some waders. It would only have taken me a couple of minutes to get to the shore, had I not been distracted by a Kikau or Western Wattled Honeyeater foraging in a tree. There weren't many waders around, but I did manage to spot a Pacific Golden Plover, a couple of Greater Crested Terns and a Wandering Tattler – that one was new for me as I had failed to locate one in New Zealand. Tattlers are pretty dull-looking waders if I'm being perfectly honest; apologies to the birders out there, but it's difficult to name other waders that 'are quite as dull as tattlers.

Nonetheless, I was happy to find a new wader species for my bird list. Waders are only second to bats in my heart. All my scanning of the shore and photographing piqued the interest of a group of kids who, in turn, attracted my attention because they wanted me to photograph them. I took some photos of them having fun in the sea, showed them, and they seemed very happy. Internally, I wondered how I would get the pictures to them, but that didn't seem to be an issue for them – viewing the shots on the back of my camera seemed sufficient. I also lent them my binoculars so they could try them out. Once we'd all said goodbye to each other, I started heading back to the city, stopping to photograph a miserable-looking hermit crab that was using a plastic film canister as a shell. I took some photos of it before a kid showed up and threw the crab out of its 'shell', probably giving it a decent chance of either finding a new one or getting eaten. I guess we'll never know, but I wasn't too hopeful.

Back at the lodge, I went out for some birding in the nearby forest. Like most Pacific islands, Viti Levu has relatively low diversity, but its uniqueness makes it special. It was my first time birding in a tropical forest on my own, and the experience highlighted that I wasn't very good at spotting birds. I didn't know any of the calls or songs, but I did try my best to inspect every movement in the vegetation. It paid off well, as I ended up finding bird species such as the Slaty Monarch, Fiji Bush-warbler and, of course, the magnificent Golden Dove. The colours of this pigeon are simply stunning, more than enough to make one reconsider the idea that pigeons are dull-looking birds! Several species of doves in the Pacific have mixes of pink, bright green, yellow and white plumage, creating birds that look like they've stepped out of a child's colour drawing. The forest too was gorgeous. Further away from the lodge, the area felt a lot wilder – even though there were still clearly marked trails and even pools that people could swim in. The signs indicated that there were shrimps in them. Should anyone decide to do a Big Shrimp Year, I'd imagine this would be a decent place to start. (I have no idea of what I'm talking about, and I could probably say the same for bats.)

I kept seeing shadows flying over and away as I walked the trail; they seemed to be flying towards the denser parts of the forest, almost without a sound. I wasn't sure what they were. My first thought was some kind of raptor, but after seeing four or five, I thought this was unlikely. Eventually, I saw one flying away, right in front of me, and I realised those were *Pteropus samoensis* in their typical habitat and displaying their typical ghostly behaviour. While I was happy with my identification from the Presidential Palace, I felt that this sighting was more representative of the species. This species has become very shy because of hunting, and even though I would see dozens of other fruit bat species throughout my journey, none of them behaved anything like these ones.

The second location I wanted to visit specifically to see bats was a cave, the only known roost of the Fijian Long-tailed Fruit Bat *Notopteris macdonaldii* on the island. Unlike the two flying fox species I'd seen, this fruit bat roosts in caves. To get there, I asked for some advice from the lodge's staff once more. They told me to get a bus to Vunidawa, about 50 km north of Suva. From there, I should have been able to find a bus to take me to the cave, but I ended up getting a taxi, mainly to make sure I got to the right place. As I arrived, one thing was immediately apparent: going back would be much more of a challenge, because I couldn't just stumble across a taxi in that tiny village. Anyway, I was there for the bats, so that's what I decided to focus on. I tried to find someone who spoke some English, which turned out to be far easier than expected – I met a couple of guys working on a house under construction. They went to get the village chief so he could show me the cave. The chief spoke a fair bit of English too, so I was able to ask some questions to understand how this cave was used and how the villagers perceived the bats. He explained that there were three species of bats in the cave, which surprised me because it's believed to only be home to one,

One of many Pacific Flying Foxes *Pteropus tonganus* sighted on Suva.

Notopteris. As we went into the cave, he also explained that they collect a couple of hundred bats for feasts a few times a year. When we finally got to the chamber containing the bats, I tried to estimate their number, but they were very high up on the walls and taking a decent picture would have been impossible. Evidently, they numbered in the thousands – but was it closer to five or fifty thousand? I couldn't tell. I had my bat recorder, which was going a bit crazy with all the typical chatter inside any bat roost. My hope was to record a different species, the Pacific Sheath-tailed Bat *Emballonura semicaudata*, one thought to be extinct on the island.

When I asked the chief how many bats he thought were in there, he essentially said he believed there was an infinite number of them. I decided to seek the advice of Dave Waldien, an American bat researcher with extensive experience in the Pacific islands (among other places). Talking to him about what I saw, it became apparent that the population was declining: a far greater area used to be covered by these bats. If the area has reduced, it is likely because the population has dropped significantly. The chief also showed me the pipes running through the cave. Those pipes are essential to the village because they provide their supply of fresh water. Additionally, the villagers also use the cave as shelter in case of cyclones and other extreme weather events. Clearly, this cave plays a major role in their lives, so closing it off is not a solution. However, it's equally clear that something needs to be done to protect those bats, and that will have to be done in partnership with the villagers. As I was trying to estimate the number of bats, I noticed a couple of smaller bat species flying very high up, near the ceiling, with a flight behaviour that I doubted could belong to a fruit bat; I was pretty much convinced this was *E. semicaudata*, the species believed to have been extirpated from the island. Again, after consulting Dave's expertise, it appeared that they had been claimed to be extinct after relatively little research, which led him to believe that I had indeed seen them. The Pacific Sheath-tailed Bat is a small insectivorous bat, the first one from the family Emballonuridae that I had seen on my journey. In many ways, this species could be considered the quintessential Pacific island bat, as it is widely distributed across the south-western Pacific islands – a very unusual distribution for an insectivorous bat. I would have liked to make recordings of that species, but unfortunately they were simply too far away to be picked up by the recorder.

I'd now seen five species, of which four were in Fiji. The Big Bat Year was off to a slow start but not a bad one either.

As we left the cave I thought again about the chief saying there were three species within, and I figured they must have counted the swiftlets as a bat species. Think about it: they live in caves, feed on insects and echolocate, producing similar clicks to the chatter of *Notopteris*. From a distance, there isn't much to differentiate them from bats. The other two were likely real bats. He said there was one larger one, *Notopteris*, and one smaller one, reaffirming my theory with the idea that they also regularly spotted a smaller bat species. Because of the limited diversity on the island, it's unlikely to be anything else. I gave the chief some money to thank him for the guided tour and headed for the bus stop. After about twenty minutes or so, there were still no buses. The locals told me I'd have to wait another twenty minutes, but I was already sceptical of the times I'd been given. Two kids came to join me, on their way back from a fishing excursion to a nearby stream. They'd caught a few small fish as well as a small crab that they were very proud to show me. Our conversations were difficult because I didn't speak Fijian, and they didn't speak English, but we managed to understand each other regardless. They seemed particularly thrilled to have caught a crab – and to be fair, I was quite happy to see a freshwater crab too.

As no buses appeared on the horizon and I was getting concerned about returning to the lodge, I asked the locals again if they could help. That's when a tourist bus appeared, seemingly out of nowhere – at that point I'd been waiting for well over an hour. The locals asked the driver if I could hop on the bus; they usually don't take passengers who haven't booked in advance, but they kindly accepted. They had no seats left, so I sat on the stairs, but I had a way home, and that's all that mattered. I thanked the locals for their precious help. I was on the way to Colo-I-Suva again, my head full of thoughts of how complex bat conservation really is.

Guadalcanal, Solomon Islands, January 2019

On the way to the homestay, Alistair, the owner, and Richard, the driver, explained to me the basics of the pidgin English they use and some basic information on the island history. I had to ask what the Solomonese did during World War II, because most films only show the Japanese and Americans. According to Alistair, the Solomonese either fled to the mountains or were spies for the Americans but didn't actively participate in the armed conflict. I thought I should use some of my time on the island to learn more about World War II on Guadalcanal, especially given that none of the major films on the Pacific War were actually shot in the Solomon Islands. For example, *The Thin Red Line* was shot in Tropical North Queensland, which is made evident by the Curtain Tree Fig and the bats. I never suspected my knowledge of bats would help me identify film-shooting locations! As it turns out, it's as annoying to realise that the bats indicate a different shooting location as films using bird sounds from another continent or during the wrong season, for example European swifts flying above an English town covered in snow. I guess I can imagine how much more convenient it would have been to shoot the film in Australia rather than on Guadalcanal, though.

Back to bats now, because I wasn't there for war-related stuff – well, only partially. I still had bats to find and no idea how to do it. While I was talking to Alistair, he seemed very interested in my unusual quest, and he told me he could show me the bats they regularly see around the homestay. As we got there, I saw that it consisted of three elevated cabins with a terrace and an unbeatable view of the ocean. The sight of mosquito nets around the beds made me happy, given this was one of the countries where malaria could be a severe issue. Tablets only protect against one of the parasites causing malaria – and in the Solomon Islands, that one is actually less prevalent, making pills far less effective. As a result, I had to do everything I could to stop myself from being bitten—that meant mosquito spray all the time, long sleeves, gloves at night, etc. I was surprised that all the native vegetation around the homestay was gone; it was just coconut and banana trees. It was very naïve of me because habitat loss is one of the biggest threats on any island, and it usually starts at the shore and goes up the mountain as the need increases. I was immediately a bit disappointed because I thought I wouldn't get to see many bats, and I started trying to find a way to get myself to the forest later on in the trip – initially, I'd only planned to lodge at this homestay. I was hoping I could make day trips from it.

I spent most of my evenings walking around the homestay and recording bats. On the first night alone, I recorded five species! That was pretty incredible, and put my doubts to rest. They were all common species but their taxonomy isn't very clear, so they could well be split from the Australian populations one day, making that evening one of the most prolific in terms of recorded species on my entire world tour. I was not expecting that when I arrived at the homestay. The species I recorded, thereby doubling the size of the list, were: New Guinea Pipistrelle *Pipistrellus angulatus*, Great Bent-winged Bat *Miniopterus tristis*, Maluku Myotis *Myotis moluccarum*, Dark Sheath-tailed Bat *Mosia nigrescens* and Bare-rumped Sheath-tailed Bat *Saccolaimus saccolaimus*. I was rather happy to record the *Myotis* as I wasn't sure exactly where to look for it. Like the *Miniopterus* species, its taxonomic status is very much unresolved right now. I mostly identified the calls based on the excellent publication written by Kyle Armstrong and Michael Pennay, two Australian bat researchers, as it covers exactly what is known and what isn't. In a world of scientific publishing that tends to focus on findings, it's nice to see this kind of work that lists what they did find as well as everything they couldn't make sense of. The authors know that this information could prove helpful to others. The fact is that identifying bats based on echolocation is difficult; it gets slightly less tricky when literature is abundant, as is the case in Europe. Even when the diversity of echolocating species is relatively low, such as on the larger Pacific islands, there has been so little research that it remains problematic.

The list of species I decided to keep for my Big Bat Year list for the Solomon Islands (and elsewhere for that matter) is a cut-down version of my acoustic identification attempts list. I had spent a long time wondering how I would count acoustic records, and I decided that if there wasn't any reasonable doubt that it could be something else, I could count it. It wasn't a strict rule with a harsh cut-off, but it was one I was comfortable enforcing, as well as one I could actually implement. It's complicated to quantify the certainty of identification. However, *Mosia nigrescens* was one species I had no issues whatsoever identifying, given its distinctive multi-harmonic calls. I didn't know the species very well, so when Alistair told me that his neighbour was sometimes finding a few bats hanging underneath a leaf, I didn't know what it could be. They tried to find them again for me, but no joy. The more I think about it, the more the behaviour they described matches *Mosia*. They tend to roost in small groups, all piled up and hanging on the underside of a leaf. I'll never know for certain, unless I go back.

One evening, while walking around the banana plantation, I found three pteropodid species, which was also somewhat unexpected, but bananas and coconuts do tend to attract these flying puppies. In fact, that's the first thing

Alistair recommended I check out, as he saw bats in the coconut trees almost every night he was out. The species I found were the Admiralty Flying Fox *Pteropus admiralitatum*, a reasonably common Solomonese endemic; the Dwarf Flying Fox *Pteropus woodfordi*, a less common, tiny, Solomonese endemic; and the Lesser Long-tongued Blossom Bat *Macroglossus minimus* (that's a mouthful), a tiny nectar-feeding bat that packs a powerful bite – as I'd discover on my evening of trapping. I'd photographed the Blossom Bat feeding on a banana flower, but I wanted to catch it to verify the identification. I'd never handled anything larger than a *Myotis myotis*, a 30–35 g bat that can be pretty keen to bite. That one feeds on carabids (a family of beetles), and a 16 g nectar-feeding bat didn't seem like such a danger. So I set up my net – with great difficulty, as sourcing poles turned out to be rather tricky. We ended up having to hold the net while trying to see when the bats would fly into it. Overall, I'd say this was a terrible idea, and it shows how little experience I'd had of trapping on my own. The *Pteropus* species were feeding high up, on banana flowers and coconut blossom. Luckily, it only took a minute or so for the bat to land in the net. But it took me a good five minutes to get it out because of how feisty it was. The more it kept fighting me, the more entangled it became.

As I tried to hold it down to extract it, the bat bit right through my leather gloves as if they weren't there. It was a genuinely painful bite, and to this day, it is still the most painful bat bite I've ever endured – and the most stressful, because I had no way of getting post-exposure rabies jabs or anything like that. It's no wonder I never told my parents about this bite that was worrying me quite a lot. In fact, they may even only hear about it when reading these lines. No such attempts would be repeated further along the way. The question that remains is why on earth a nectar-feeding bat needs a stronger bite than a bat twice its size that crushes insects with notoriously thick shells. Possibly it's because their teeth aren't essential to their diet (unlike snakes for example), and thus they have no reason to hold back on their defence. As traumatic as my experience was, this bite was a defence mechanism; it always is with bats. They don't like being handled and they fight back.

After only a few days, I'd already found eight species near the homestay. That wasn't bad at all, and in line with my initial goal of finding two species a day. This was definitely far more successful than my Fijian and New Zealand stays, but this was also the first location with a decent diversity of bats, so it made sense. I wanted to try to find more, however, so I asked Alistair if he knew of any likely spots. He said he knew of a cave that had bats. When we tried to visit it, we couldn't find the person who had the key and getting there seemed very complicated. I had a feeling they were pretty reluctant to have a 'tourist' visiting the cave, which I can understand. When I asked Alistair if the forest was accessible from the homestay, he told me that it wasn't and that finding roads and getting to the good parts would be almost impossible. However, he suggested that I stay at a lodge up on the mountain surrounded by native forest. He knew the owner, so he was able to quickly arrange my stay there, and transport to get me there from the homestay.

I learned a lot of things that day. First, I realised that it would have been impossible to plan absolutely everything ahead of time; bats can be so unpredictable. Staying all week at the mountain lodge would have been a terrible idea, for example, as of the eight species I got from Alistair's place, I only found three up there. Disturbed habitats can be better environments for bats than for birds, so relying on information from birding trips also had its limitations. Second, I learned to ask for help. I've struggled with this for the longest time. I could always figure most things out on my own without needing any assistance. But for a trip like that, for a marathon like that, I would need help, and I would have to learn to ask for it and be grateful when it was given to me. Alistair became the first of many to help me in my quest for bats, despite knowing very little about bats himself.

The habitat around the mountain lodge couldn't have been more different from Alistair's homestay. Well, maybe if it had been a desert. But this was clearly secondary forest, lush and noisy from all the birds inside. Just beyond this bit of forest, the primary forest was visible too. This part was close to the logging road and had lost most of its large trees already. The birdlife was incredible nonetheless. Bats, on the other hand, didn't want to play ball. The densities appeared to be much lower than they were down by the coast and I failed to find a new species. Perhaps staying longer would have changed that; I guess we'll never know.

After a quick glance at the day trips offered by the lodge, I noticed one included a waterfall hike with a cave. I was quick to ask if the cave had bats, and as soon as they said yes, I asked to leave as soon as possible. We left shortly after that, as I was told this would be an easy one-hour hike each way. It wasn't anywhere near that. It was over three hours each way because it involved crossing a powerful and deep stream between 15 and 20 times; I sort of lost count at one point as I was more worried about not dropping my camera than anything else. My two local guides had no issues with this trail whatsoever; I can definitely believe they could do it in one hour. There's no way I could have. I had to remove my walking shoes every time we crossed, and while I tried crossing barefoot in the beginning, it was too painful, so one of the locals lent me his flip-flops. They helped with the sharp rocks, but then they also kept being ripped off my toes because of the strong current. This was an exhausting hike that drained all the energy I had. I thought about turning back a few times. I became cranky and somewhat rude. I didn't want to take time to look at birds; we saw a flying fox swooping away that I'll never get to identify. When we got to the waterfall at the end, we had to wade up to our shoulders in a pool to get to the cave. One of the guides took my camera and carried it well above his head, so I could focus on not dying, essentially.

The species that was in the cave? Was it worth it? Heck yes! It was the Solomon's Naked-backed Fruit Bat *Dobsonia inermis*, a rare endemic species from

Outdoor World War II museum on Guadalcanal, Solomon Islands.

the Solomon Islands, one I'd never read about in reports and that was rarely mentioned in the literature. If this is what it takes to see a roost, I can easily understand why. Adventures like this one, that involved taking a leap of faith to potentially see a rare bat, are why the Big Bat Year excited me so much. The walk there may have been hell but in the end, it was worth it. The walk back wasn't anywhere near as bad, surprisingly, but seeing the car waiting definitely put a smile back on my face.

Island bats

Islands formed a considerable part of the first half of the Big Bat Year. They often exhibit low diversity in terms of actual number of species, but they are often unique. Islands also come with their unique challenges. The limited area means that most threats can quickly spread to the entirety of the island. The majority of islands are located in the Pacific, coincidentally putting them on the path of extreme weather events.[1] The underlying threat of severe storms makes species vulnerable to the all too common danger of habitat loss. In turn, this leads many species, including bats, to forage closer to urbanised areas, therefore exposing them to hunting and the threat of invasive species.[2]

The harsh weather does not make islands the paradises for bats that they are for beach lovers. Yet 230 bat species are only found on islands, with a further 330 also inhabiting continental land masses. Of those 230 island species, about 75 are single-island endemics.[3] On many of those islands, bats are also the only native land mammals as they are the only ones capable of flight, making them more able to disperse over vast distances. For many taxa, the highest diversity is found in Ecuador and Colombia. That's true for amphibians, birds, orchids and likely many more. However, when it comes to bats, there is a country with 20% more species: Indonesia. With its 17,500 islands, it is no surprise that the country has over 235 described species and many more waiting to be discovered or described, such as the *Rhinolophus species novo* that I found in West Papua.

Unique genera or even families such as *Pteralopex*/*Mirimiri* (the Solomon Islands and Fiji), Myzopodidae (Madagascar) and Mystacinidae (New Zealand) are what make islands incredible study sites for speciation and evolution in general.

Sadly, islands tell many tales of conservation shortcomings or even failures. Big South Cape in New Zealand, Christmas Island, Lord Howe Island, and Tasmania in Australia have suffered many extinctions in recent times. Most often, only the consequences of extinctions, i.e. the subsequent absence of a given species, have been documented in detail. However, the stories from Big South Cape Island and Christmas Island are an exception.

The species were observed to be declining, but nothing was done in time to save them. On the former, translocation attempts were made for the Bush Wren, South Island Snipe and South Island Saddleback. Only the latter species was successfully translocated in the mid-1960s, and it still survives to this day on several offshore islands around the South Island of New Zealand. The bats were not translocated as they were believed to belong to the same species as the New Zealand Lesser Short-tailed Bat *Mystacina tuberculata*. *Mystacina robusta* became extinct shortly after as it was very vulnerable to the newly arrived rats.

Christmas Island tells a similar story of a decline witnessed by researchers and authorities alike. In the mid-1990s, a range reduction and population decline had been noted. Between 1994 and 2008, it's estimated that 99% of the population had disappeared. It is now considered extinct. This rapid decline in both the Christmas Island Pipistrelle *Pipistrellus murrayi* and *Mystacina robusta* is reminiscent of the rapid extinction of the Steller's Sea Cow and the Dodo. Both of them were driven to extinction well within a lifetime.

'Too little, too late' is probably the best way to sum up these particular tales. However, much was learned in their aftermath, and New Zealand now has numerous pest-free islands and mainland sanctuaries where many species of birds, reptiles and bats are thriving. Let's hope that the lessons learned allow us to protect some of the rarest non-extinct bats in the world, such as *Pteropus livingstonii* (Comores), *Mirimiri acrodonta* (Fiji) and *Coleura seychellensis* (Seychelles). In conclusion, while islands are often considered jewels in terms of evolutionarily distinct species, those jewels are also fragile, and we must protect them.

Notes

1 Pierson, E.D. and Rainey, W.E. (1992, July) The biology of flying foxes of the genus Pteropus: a review. In Don Wilson and Gary L. Graham (eds) *Pacific Island Flying Foxes: Proceedings of an International Conservation Conference*, vol. 90, pp. 1–17. Washington, DC: US Department of the Interior Fish and Wildlife Service.

2 Palmeirim, J.M., Champion, A., Naikatini, A., Niukula, J., Tuiwawa, M., Fisher, M., Yabaki-Gounder, M., Thorsteinsdóttir, S., Qalovaki, S. and Dunn, T. (2007) Distribution, status and conservation of the bats of the Fiji Islands. *Oryx* 41 (4): 509–19. https://doi.org/10.1017/S0030605307004036

3 Jones, K.E., Mickleburgh, S.P., Sechrest, W. and Walsh, A.L. (2010) Global Overview of the conservation of island bats: importance, challenges, and opportunities. In Theodore H. Fleming and Paul A. Racey (eds) *Island Bats: Evolution, Ecology, and Conservation*, pp. 496–530. University of Chicago Press.

Grande Terre, New Caledonia, January 2019

Once again, I had just landed in Auckland as there were no direct flights between Honiara and Nouméa. I landed at 1 am, when everything inside the airport was closed, including the transfer desk. All I could do was wait until things opened again at five o'clock. Along with dozens of other passengers, I found a spot in one of the corridors and waited. Sitting on a hard floor, watching Netflix in the middle of the night, wasn't exactly ideal after a week spent in the scorching heat of the Solomon Islands. Instead, I tried to find a quiet area with some seats to sleep on, lying down. After roaming around the only corridors we had access to, I found a nice sheltered spot with three seats. This was far more comfortable than sitting on a corridor floor. A couple of hours went by, and finally, the airport opened again. I headed for the lounge, where I intended to spend the remainder of my 13-hour stay. There, I could take a well-needed shower, have some great food and get some work done.

After this day spent enjoying the airport facilities, it was time for me to get back to finding bats, in New Caledonia this time. In a way, I was relieved to be headed to a country where I spoke the language, but I wasn't sure what to expect. How much like France would this be? Upon landing in Nouméa, I was happy to be able to use my Belgian ID card instead of my passport to get through immigration. It didn't actually make much of a difference compared to using my passport, but somehow it just made the process less stressful. I had hired a guide, Isabelle, who was waiting for me outside the airport. I could only afford one and a half days of her services, as I quickly discovered that, unlike on neighbouring islands, everything was prohibitively expensive. Isabelle kindly explained to me why she had to charge this much to be able to make a living, by detailing the ridiculous economic context of the nation.

Isabelle had agreed to help me in my quest for bats endemic on Grande Terre. However, our first stop felt like a long shot to me. I told Isabelle we were going to try to find a species not seen or heard since 2002. We headed for the hills, the site of its last known location. Upon arrival, we still had a couple of hours of sunlight at our disposal, so we did some birding in the rainforest of Mount Koghis. Here there was a pocket of temperate rainforest isolated in a landscape dominated by exotic pines and shrubs. It reminded me a lot of the forests of New Zealand that I had fallen in love with in 2016, and then again a few weeks before finding myself in New Caledonia. There aren't many areas of temperate rainforest on our planet, let alone ones filled with exotic-looking ferns. The birdlife was exciting, but we didn't see any of the main species people travel to this part of the world for. But birding wasn't our goal that day. Our target was the New Caledonia Long-eared Bat *Nyctophilus nebulosus*, a species of which no pictures of live individuals exist; nothing is known about its biology, roosting habits or feeding behaviour. It's one of those species only known from a handful of museum specimens.

Luckily, an expedition in 2002 was successful in locating a few individuals, so I had somewhat current information to go by. I had no intention of trapping there, as that would have required permits that would have taken me months to obtain. And that was ignoring the fact that I was unlikely to even get them without any university affiliation, which is often a requirement. Most importantly perhaps, I didn't believe I'd find the bat. It seemed that the odds of me finding one were minute. All I had was my bat recorder that had helped me score seven species since I started the voyage. It was a slow start, but the quality of the species and the experience more than made up for it. Previous Big Years had skipped most of the Pacific islands because they're 'poor value' in terms of the new species per day ratio. This ratio is all that matters when

Gorgeous, yet non-native landscape. This should have been rainforest.

New Caledonia Long-tailed Fruit Bat *Notopteris neocaledonicus*.

one tries to break a record. However, I had more than that in mind; I was in it for the experience, the thrill of chasing rare species, no matter how long it took – the joy of finding something special, something that hadn't been seen in 17 years.

At 20:27, after a long wait, albeit made more bearable by a gorgeous sunset over an equally breathtaking landscape, some bat calls showed up on my recorder. The shape was unlike anything I'd seen so far, and I had seen the other echolocating bat species that are known in the area. It had to be *Nyctophilus nebulosus*! Steep calls, ending at around 25 kHz with pronounced harmonics, sort of like a mix between the *Plecotus* and the *Myotis* that I knew well from back home. The fifteenth species of my trip was in the bag, and it already had a place on the podium of the whole year's highlights. It would be a hard one to beat. When I'd left Honiara, it was with only nine species added to the list in the Solomon Islands. It was fewer than I had anticipated and I felt very disappointed. However, those negative feelings were immediately drowned by the happiness of rediscovering a species thought to be extinct. And no one had made sound recordings of it ever before! At this point, you might be asking yourself how I identified it if those were the first ever recordings. Over in Australia, the genus *Nyctophilus* is quite diversified, and there is considerable overlap between the calls of a few species. It would seem very unlikely for the New Caledonian members of the genus not to look like their Aussie cousins, and indeed they did.

For the rest of the stay, I'd got in touch with Lucie, one of Isabelle's friends. She was able to help out with the caves in the north-west of the island that I wanted to see. Getting in touch with landowners to secure permission to access caves that I'd found vague GPS coordinates for was a challenging task. Without Isabelle and Lucie, it's unlikely that I would even have been able to add another three species to my list. Those species were the Ornate Flying Fox *Pteropus ornatus*, New Caledonia Flying Fox *Pteropus vetulus* (or *vetula* depending on whether or not you think grammatical mistakes should be corrected in scientific names) and New Caledonia Long-tailed Fruit Bat *Notopteris neocaledonicus*, of which I was able to get some good photographs. The latter species has only rarely been photographed alive and good photos always help with conservation outreach. I'd seen all the species found on Grande Terre. Only the Loyalty Long-fingered Bat *Miniopterus robustior*, restricted to the Loyalty Islands, had escaped me – I hadn't tried for it, as I didn't feel it was worth the time and money. New Caledonia is definitely not an easy place to visit for bats but it can be incredibly rewarding, with the right people helping.

Queensland, Australia, January 2019

Less than four hours after landing in Australia, I was in the field again after a quick stop to drop my backpack in Brisbane. It felt weird to be back in a large city after a few weeks spent mostly on small islands. Vanessa Gorecki was doing her PhD at Queensland University of Technology on the use of human-made structures by bats in urban areas, mainly concrete bridges and culverts. As she was following a number of sites in Brisbane, she offered to help me in my quest to find more bats. She came to pick me up at my hostel, and then we were off to the Botanical Gardens, where we picked up Luke Hogan, a bat acoustics expert who occasionally tests out kit for Titley Scientific. He was the perfect person to introduce me to Australian bat calls. It's one thing to be able to navigate the calls of the handful of species present on a small island; it's another thing entirely in Australia. Australia has over 70 species of echolocating bats – microbats, as Aussies still very often call them over there. The denomination of micro/macrobat was abandoned around 2013 following a publication showing that families such as Hipposideridae (Old World leaf-nosed bats) and Rhinolophidae (horseshoe bats) were much more closely related to 'macrobats' – i.e. Pteropodidae (flying foxes and allies) – than they were to Vespertilionidae (vesper bats) and Phyllostomidae (New World leaf-nosed bats). This has several implications from an evolutionary perspective; the main one of interest to me is the conclusion that echolocation either evolved twice or it only evolved once, subsequently disappearing in Pteropodidae, and the related families turned the concept on its head (more on that on bat evolution later in the book).

Back to the culverts, then! We stopped on the side of a road near Mount Coot-tha, where Vanessa had found both the Large-footed Myotis *Myotis macropus* and the Australasian Long-fingered Bat *Miniopterus orianae* in the past. We only found one instance of the latter, as the culverts had surprisingly few bats. Not really an issue for me, but Vanessa was a bit frustrated. As we had some time left before it got dark, we took the opportunity to exchange information on our respective projects. We then drove to a different part of the reserve and started walking around the area with our bat recorders turned on. Because I had been

Urban flying foxes in Brisbane.

travelling all day, with no opportunity to recharge my phone, I was running low on battery. This has to be the main downside to using your smartphone as a platform for your bat recorder, something I would learn time and again over the following months. I was hoping it would at least last me the first hour of our bat walk. Luke had brought along his Anabat Walkabout, a device I'd heard about but not seen in person. I wasn't too convinced it would work for me at the time, but after a whole year of travelling the world listening to bats, I decided it definitely was the right tool for me and bought one in May 2020. At the time of writing these lines, I can tell you I truly love this thing.

Going back to late January 2019, though, I was curious to see how it performed compared to the one I was using, as it costs a fifth of the price of the Walkabout. Our first bat was a species of *Scotorepens* (a broad-nosed bat), which isn't the most exciting species to start with because, in this part of the country, there are two species present that are virtually indistinguishable from each other based on their echolocation. My first species in Australia could therefore not go on the Big Bat Year list. The second species, however, was Gould's Wattled Bat *Chalinolobus gouldii*, which I recognised quite easily because it looked rather similar to the other *Chalinolobus* I'd recorded in New Zealand and New Caledonia. This brought a slight feeling of relief, as I'd started to think maybe touring Australia with a bat recorder wasn't the easiest thing I'd do all year.

A bit further down the path, it became rather noisy; not on the bat recorders, we could hear these bats without them. They were flying foxes – Black Flying Foxes *Pteropus alecto*, to be precise. They were feeding on a fig tree. In the dark, they did indeed look rather black, but then that didn't mean much. Our ears would turn out to be as useful as the recorders for the following two species as well: Yellow-bellied Sheath-tailed Bat *Saccolaimus flaviventris* and White-striped Free-tailed Bat *Austronomus australis*. Both are species that can be heard with the naked ear. *Austronomus* is an especially loud bat and turned out to be rather inquisitive too – shining a torch to the sky attracted the bat, and we could see the white stripes from which it gets its name. Luke and Vanessa both knew the area very well, so they knew exactly which locations would have which species. We headed towards a series of small ponds that they predicted would be good for sighting the Large-footed Bat *Myotis macropus*, also the only *Myotis* in the country. There were lots of Cane Toads when we got there but sadly, no *Myotis*. The ponds had quite a lot of surface vegetation, which is a challenge for any trawling bat species. The leaves on the surface of the water don't reflect bat echolocation calls as well as the water surface itself, so they end up confusing the bats. The name *macropus* refers to the large feet of this bat species, which is a characteristic feature of all trawling *Myotis* species such as Daubenton's Bat *Myotis daubentonii*, the Pond Bat *Myotis dasycneme* and the Mexican Fishing Bat *Myotis vivesi*, but also of other trawling bat species such as the bulldog bats of the Americas (*Noctilio* spp.).

Next I flew to Cairns to meet a childhood friend of mine. Oscar's parents and mine have been friends for a long time, so we've spent a fair bit of time together over the years, but the age gap meant we weren't really close. Despite this, when we realised we'd be in the same place at the same time, we wanted to plan something. We eventually decided to rent a campervan for a week and tour Tropical North Queensland. Before leaving Cairns on our road trip, I visited the library, not for the books but for bats. There's a well-known roost there of Spectacled Flying Foxes *Pteropus conspicillatus*. It's notable because it's an urban roost, and those tend to be popular, but also because Cairns authorities regularly try to get rid of it, citing nuisance to the neighbours, often leading to significant backlash from wildlife enthusiasts and bat workers alike. In addition to the flying foxes, I went on the search for other bats as well, in the evening. I headed up to the harbour because it looked rather green on the satellite view, and indeed there were a fair few bats around, such as the Little Long-fingered Bat *Miniopterus australis* and the Greater Northern Free-tailed Bat *Chaerephon jobensis*. Only the latter was new for me as I'd recorded *M. australis*, or what is still considered the same species, in the Solomon Islands and in New Caledonia. However, it was mainly the birdlife that caught my interest on that stroll, with Water Thick-knees walking around on the footpaths, pelicans off the shore and waders calling in the night. I had never done any nocturnal birding along the seashore, and I must say, it was delightful. This beautiful evening took a turn for the worse, however, when on the way back to my hostel I had the nasty surprise of realising that my EchoMeter had died. I suspect the heat and humidity of outside followed by the coolness of the air conditioning in the room gave it a shock. I could hardly sleep, given I was sure this meant the end of my trip, or at least the failure of a large portion of it. I did not expect it would be easy for me to find a replacement recorder on short notice. Panicked, I tweeted at the manufacturer, who promptly responded by suggesting I got in touch with their support staff, something I should have thought of immediately. I emailed them before going to bed, not expecting a reply within a few days. Once again, their efficacy exceeded my expectations. They'd reached out to their local supplier and I was told to send it off to them for examination, repairs or replacement.

The next day, Oscar and I were off in our rented campervan on a tour of Tropical North Queensland. It seemed perhaps a bit ambitious to do given we barely knew each other, but it felt like an opportunity not to be missed. Another challenge we hadn't foreseen was the consequences of the recent cyclone. Most roads out of Cairns were flooded. The blocked roads meant we couldn't drive south or west, and we were limited in our options going north. North is where we decided to head, because that's where the true Queensland tropics start.

We were getting closer to the rainforest. The vegetation was getting lusher and lusher, and we were getting wetter and wetter. I had high hopes of finding bats there, but this would prove rather difficult given my recorder wasn't working properly. On the way there, we stopped along the coast as I'd been given coordinates to look for the Coastal Sheath-tailed Bat *Taphozous australis*. Upon arrival, we were greeted by various signs warning of the presence of crocodiles and jellyfish. I was significantly more worried about the former than the latter. Back in Brisbane, I'd been warned to remain 20 metres away from the water as that's usually how far Saltwater Crocodiles can jump. While looking for the sea cave, Oscar and I were very careful to check the beach to avoid any unpleasant surprises! The cave itself was challenging to access, as it appeared to be permanently flooded with various debris brought in by the sea, which created a dam. We tried to dig a canal to drain it and access the cave, but to no avail. There were no bats to be seen either. This hadn't been a very successful stop.

After that, our first destination was Julatten, in a rather basic campground, but as we had everything we needed in the van, it made no difference to us. Without my recorder, I was resigned to looking for geckos around the facilities, which I 'wasn't able to identify. Not having a bat recorder also meant I could have an early night. The main target of this trip was the Wet Tropics, which we were still a bit south of. On our way north to Cape Tribulation the next day, we stopped for a dead echidna on the road. Echidnas are egg-laying mammals, like Platypus with spikes on their backs and a long snout. They're unique-looking animals that I had been hoping to see alive. Unfortunately, it's not uncommon to see them on roads, having fallen victims to traffic. As we reached our destination, it became apparent where the name 'Wet Tropics' came from; it was wet, very wet and rather hot too. This rainforest didn't look like what I'd seen in Fiji or the Solomon Islands. I'm far from being a botany expert, as I can barely tell two trees apart, but I could see that there were differences.

My odds of finding new bats were still very much hindered by the fact that I was without a functioning recorder, but there was one species I was hopeful for, Queensland Tube-nosed Fruit Bat *Nyctimene robinsoni*. This strictly nocturnal fruit bat is part of a larger genus, distributed across Melanesia, from Seram to the Solomon Islands. They are challenging to find as they are often rare, but in Australia *N. robinsoni* can be spotted relatively easily in the right areas. What makes its sighting even easier is the absence of similar species. This meant that when I saw a smallish fruit bat fly across the road shortly after dusk, I knew exactly what it was. This bat had a slow flight, somewhat typical of fruit bats, but had much broader wings than flying foxes. Those broader wings, thanks to the increased manoeuvrability, help it fly through the forest as opposed to travelling above it as most flying foxes do. Unfortunately, this brief sighting didn't allow me to spot the beautiful Yoda face these bats have. I hoped Tolga Bat Hospital, a bat rescue centre I'd visit a few days later, would have one in care to see up close.

The rest of the evening was relatively quiet, with some time spent in the pool, which was nice and refreshing. While spotlighting for mammals and hopefully for bats, I managed to attract and confuse a massive Atlas Moth. I'd read about them as they are the largest moth on the planet and are somewhat frequently kept in captivity – but I had no idea I'd ever see one, let alone have one dive down towards me. I'm usually not scared of insects but seeing one the size of a dinner plate diving right at me was mildly disturbing. I took a few photos and then turned off the light so that it could go back to its mothy business.

For the next couple of days, Oscar and I kept ourselves busy with various animal species, visiting whatever sites we were able to, given the weather conditions. Some of the places we'd read about were inaccessible due to flooded roads. We did try our luck on a couple of them but figured it wasn't worth the risk. We decided to visit Mareeba for its famously tame rock-wallabies. There too, the signs of the intense rainstorms were evident and there was even a whole chunk of the road missing, meaning we had to be extra careful and follow the workers' instructions on how and when to get through. We got to the location I'd plotted to see the Mareeba Rock-wallabies. I think in Australia, the Mareeba Rock-wallabies' fame is only second to the Quokkas from Rottnest Island (Western Australia), known for being so friendly that people can take selfies with them. While Quokkas appear rather tame by nature, the rock-wallabies have become tame because people feed them.

When we arrived, the landscape was very different from what we'd seen on the way. We drove through a dry forest, although not so dry given the recent rain; we even almost got stuck on the muddy road, but it gave way to a beautiful, somewhat lunar-looking landscape. (Well, I don't know what the moon looks like up close, and I don't know if it's got rock-wallabies'!) While

Teaching a rock-wallaby about wildlife photography.

picturesque, walking on the rocks wasn't getting us any wallabies at first. How elusive could tame, almost domesticated wallabies be? It turned out, not very. There they were, a little further down the rock formations: the famous gang of mischievous wallabies. It was immediately apparent that they would be rather challenging to photograph, not because they stayed at a distance, but because they wanted to have their faces pressed right up against our camera lenses. I'd picked my primary lens, my Nikon 300 mm f4, mainly because it allows me to photograph insects and bats up close – but it wouldn't focus *that* close. So I put it down on the rocks, took a couple of steps back and took out my phone to snap a picture of a wallaby checking out my camera gear. I've named this photo 'Being bullied into giving photography classes' because the animals were relatively bullish in their requests for food! We did not give them any. Interfering with wildlife in that way is against what I stand for. I would never feed an animal for a picture.'

Next we headed to Chillagoe, a town on the edge of the desert, because we both wanted to see the desert. It wasn't quite the arid red desert typical of the Australian Outback. For that, one would have to drive even further west, for an entire day. The vegetation was very much savannah-like but instead of lions, there were kangaroos or wallaroos (I'm not quite sure I understand the difference). It should have been a dry environment, except it had been raining a lot, so nothing was actually dry. We knew there was a cave in the area with bats but couldn't get there. Instead, we set out to find an abandoned mine. We entered without hesitation, as there was no fence or sign to be found, and spend some time photographing the various ruins and machines. We were unable to find any bats in those ruins, however, despite some of them looking like perfect potential roosting spots. Unfortunately, coming back there in the evening would have been pointless because I was still without a functioning recorder.

I'd been given vague to a cave with the Dusky leaf-nosed Bat *Hipposideros ater*, one of the few Hipposideridae in Australia and what would become my first ever member of that species on my life list, as well as the year list. The cave also happened to be quite close to the famous Etty Bay Beach, known for its roaming cassowaries. This seemed like a good spot to stop for a quick lunch before we went out looking for that bat cave. We knew it was behind a campground, but finding said campground was challenging. Once we found it, a quick chat with the owner made things a lot easier. He was definitely surprised by our request to visit the cave to photograph bats, but he was very happy to help. In fact, he also told us to spread the word about the bat cave so that other people like us would visit. Unfortunately, at the time of writing, the campground is no longer open. The cave itself was small, and the bats easy to find, even without having to actually enter the cave – they were easily visible from the outside. While most of them were of the typical brownish colour, a few of them were bright orange, which has led many people to report this species as the Orange Leaf-nosed Bat *Rhinonicteris aurantia*. In reality, however, a number of cave-roosting bat species can turn orange due to the iron concentration in their roosting environment. It takes a lot more to identify a bat than a colour.

The following afternoon, we had a quick look at the famous (there are a lot of famous things in the region) Curtain Tree Fig, featured in the opening scenes of *The Thin Red Line*. Before we knew it, it was time to meet Alan Gillanders, a local wildlife guide who'd agreed to show us the night-time wildlife in the area. Our targets with him were mainly the Large-footed Myotis *Myotis macropus*, a bat I didn't need a recorder for, and the Platypus. Luckily, both species were present on the same site, making Alan confident that we would get to see both. Before heading there, though, we made a quick stop at the Tolga Bat Hospital. Alan and Jenny, the manager, are long-time friends, and Jenny had kindly agreed to lend me their EchoMeter for a couple of days so I could find bats in the evenings too. This good news was all I was waiting for to ship my EMT to the Wildlife Acoustics retailer in Southern Australia for them to inspect and hopefully repair. At Tolga, we had a quick look at the baby flying foxes, but we were on our way soon after as we'd planned to come again the next day for the full tour. The sighting of *M. macropus* was on the underside of a concrete bridge, a common sight for many fishing bat species. The Platypus was both exciting and underwhelming at the same time. I'd say the animal itself was exciting because we all know how bizarre they are, but the sighting itself was underwhelming because there was little more to it than what looked like an occasionally diving log.

On our way back to Tolga for the full tour, we visited the garage of a friend of Alan's to identify the bats they'd found there. It was funny and slightly surprising to be called to give an expert opinion on something like this, but they seemed

Cuddle puddle of Eastern Long-eared Bats *Nyctophilus bifax*.

more confident in my identification abilities than I was. When we found the bats, however, they were beyond the shadow of a doubt the Eastern Long-eared Bat *Nyctophilus bifax*. Visiting another one of Alan's friends yielded a very cute mama pademelon (a kind of small wallaby) with a joey occasionally venturing out of the pouch, as well as a very tame Magnificent Riflebird that came to get food from my hand. We eventually got to Tolga, where Jenny explained all the work they do. As far as bat rescues go, this is as crazy as it gets. Nine hundred orphan bats had been collected that year, all cared for and released a few months later – an awe-inspiring achievement by this fantastic team of volunteers and staff. In addition to the hundreds of orphaned flying foxes, they also care for 'microbats', as they are still called there; they had a couple of *Nyctophilus* as well as an *Ozimops*. Some bats also live there permanently because they cannot be released into the wild, such as the famous (I keep telling you, everything is famous there) Lady Di, a Diadem Leaf-nosed Bat *Hipposideros diadema* that's been living there for years. One thing they didn't have was a Tube-nosed Fruit Bat. While I'd later see one up at Cape Tribulation, I wanted to see one up close, and I was hoping they'd have one in care, but the ones they did have had been released shortly before. However, during the tour, they got a call saying one was on the way, so we stayed a bit longer to wait for it to be checked in. They really are adorable bats, with their Yoda-like ears and weird nose tubes. This was yet another female that had got caught on barbed wire – by far the most common reason these bats require medical assistance.

The evenings were packed with arboreal marsupials, both diurnal – specifically tree-kangaroos – and nocturnal, such as the Striped Possum, one of the weirdest ones out there with its incredibly long fingers. Because I had a bat recorder again, albeit not mine, I was able to record some more bats, including Gould's Long-eared Bat *Nyctophilus gouldi*, the other long-eared bat in the Wet Tropics, and the Eastern Forest Bat *Vespadelus pumilus*. Alan was keen to know what features I looked at to identify those species, as the world of bat acoustics was almost entirely new to him. And what's the point of knowing things if one can't share them with others? Our evenings ended quite early, as the rain started and I truly did not want to flood Jenny's recorder.

Darwin, Australia, February 2019

I'd reached out to David Nixon, a British ecologist who regularly visits Australia on his way to and from Papua New Guinea, where he trains people to handle venomous snakes and does outreach. He was happy for me to join him when he was out and about, but he told me he wouldn't be looking for bats, only herps, mainly snakes. He has a strict no bats policy during his holidays, as it's already his day job. I checked into the same hotel as him for easier logistics, but had hired my own car. The hotel was a rather posh resort but was a welcome break – I had underestimated the importance of a bit of luxury. Over lunch, David and I discussed the wildlife in the region, the options I had, and the places he recommended I visited. I'd been in touch with a local researcher, Damian Miles, who'd also given me advice, so it was just a matter of figuring when I would go and where. One thing was certain though: David did not want to go on acoustic surveys, which I really couldn't blame him for; a holiday needs to be a holiday. David suggested I tag along on his quest to find death adders, with some birding along the way. Like bats, reptiles are easier to find at night. Combining the two was an unexpected, yet realistic, option. We managed to find the Rough-scaled Death Adder *Acanthophis rugosus*, the species David was looking for, and the birding turned out to be excellent as well. Unfortunately, without a recorder, I didn't manage to find any bats. The pace of herping also didn't seem compatible with looking for bats. As a result, David and I realised we'd have to go our separate ways if we were to have successful trips. Combining bats and reptiles wasn't too realistic after all.

The next day, after a relaxing day by the pool, I picked up a parcel at reception – my replacement bat recorder had arrived. I was genuinely impressed by the quality of service from everyone involved. As a result, I was back on the road for evenings of batting and adding more species to the BBY list'. The manufacturer wasn't able to figure out the cause of the issue, as they could find no evidence of water ingress. They ended up shipping me a brand-new one as a replacement under warranty. They were even kind enough to send me photos of the circuit board for my own interest, as I was curious to see what the insides looked like.

Breathtaking landscapes, home of the Ghost Bat.

One of the places David had mentioned was Fogg Dam, a site famous for its crocodiles. I wasn't so keen on leaving the car when on the dam and I spent the evening driving back and forth, hoping to record new bat species. The intense humidity and heat were additional good reasons to stay in the car as well. With a total of six new species added to my list, I think it's fair to call this a bat-filled site! The various *Nyctophilus* and *Taphozous* species I was able to record in the area made for some interesting identification challenges. I'd missed my opportunity to see the Ghost False-vampire Bat *Macroderma gigas* while in Queensland, but I wasn't ready to leave the country without it – so I set out to find a site in Northern Territory where the species was present. Damian recommended Lichfield National Park as there is a well-known roost there, close to a waterfall. Thankfully for the bats, the large drop between the viewing platform and the cave means that they cannot be disturbed. The scenery was jaw-dropping: a large canyon with a waterfall in the middle, all lit by a setting sun. Interestingly, despite being predators, the Ghost Bats aren't the only occupants of that cave. It's also inhabited by the Orange Leaf-nosed Bat *Rhinonicteris aurantia*. That species is much smaller than *Macroderma* and does occasionally become dinner for it. To avoid getting eaten, they have to make sure to vacate the roost by the time *Macroderma* emerge. It really makes you wonder why they would even use the same caves. I was unfortunately unable to record *Rhinonicteris*, but shortly before it started raining, a Ghost Bat flew close above my head. The distinctive large pale bat also makes noises audible without a recorder, which I wasn't really expecting.

While back in Brisbane for one night, I visited Mount Glorious, a wooded area close to the city centre, hoping to find something new, which I didn't. Shortly after, I travelled to the temperate part of Queensland. Temperate forests aren't really what one would expect to find in Queensland – and yet it's where species such as the Golden Bowerbird hide. Lamington National Park offers a range of habitats, including rather dry forests down close to the coast, all the way to the temperate rainforest at the top. Those rainforests, while somewhat similar to the ones in New Zealand, differ in many ways, with the perhaps unfortunate exception of the Short-tailed Brushtail Possum, which was introduced in New Zealand and has feasted on countless native species there. Around O'Reilly's Rainforest Retreat, I was able to record two new species of bats: the Chocolate Wattled Bat *Chalinolobus morio* (as a Belgian, how could I not like such a species), a common species in Queensland that doesn't inhabit any of the wet parts of the state, which explains why I hadn't yet recorded it; and the Golden-tipped Bat *Phoniscus papuensis*. I couldn't have hoped for a better species to finish off my Australian stop – even though for the first time since my arrival in Oceania, I was actually cold. The Golden-tipped Bat specialises in catching spiders from their webs and is known to roost on the underside of bird nests. The golden tips of its fur gave it its common name and happen to provide excellent camouflage inside bird nests too. A truly fascinating species.

Before leaving Australia, I had one more stop to make. It wasn't to find bats; it was to meet a bat researcher with whom I'd been unable to schedule a field trip. I had contacted Professor Stuart Parsons the year before to discuss a potential PhD, possibly on Ghost Bats. While this didn't come to anything and the Big Bat Year happened instead, I still wanted to meet him to discuss Australian bats and current research. This year was as much about finding people who have dedicated their lives to bat research and their conservation, as it was about finding the bats.

Echolocation

Echolocation is probably one of the most fascinating aspects of bats, at least in my opinion. In short, echolocation is a system enabling the orientation in space through the emission of (ultra)sounds and the interpretation of their echoes. Those sounds have been described as covering durations from 3 to 300 milliseconds, frequencies from 8 to well above 200 kilohertz (the audible range of a peak-performing human is 20 kHz, the average is likely closer to 15) and repetition rates of up to 160 calls per second.[1] Echolocation has played an important role in the diversification of feeding habitats, particularly in low-visibility conditions, i.e. night-time.[2]

A few general rules are followed by echolocating bats, such as the smaller the bat, the smaller the target prey, and therefore the higher the frequency. While there are many exceptions, the best examples of this correlation are found in the Rhinolophidae family where the scaling is almost perfect.[3] Researchers have found, perhaps somewhat unsurprisingly, that measurements of parts of the nose and the ears are better predictors of a species' frequency.[4]

Echolocation is a vast topic, and there are many books dedicated to it. Here are a few highlights of the incredible diversity found in bats.

Rhinolophidae and Hipposideridae use mostly constant frequency calls. They can then determine whether the object their calls are reflected by is fluttering or not, thanks to the Doppler effect. Bats compensate for the Doppler shift in their own call, and are therefore able to produce a near-perfect constant frequency. In addition to providing information on the distance and speed of their target, the changes in amplitude and frequency inform the bats on whether or not what they hit flutters, i.e. on whether or not it's a tasty snack.[5]

Murina and *Kerivoula* use very steep echolocation calls, highly adapted to cluttered environments and hunting spiders as it allows them to spot the webs. Those two genera mainly comprise tiny bats that could easily get tangled up in spiderwebs and end up as dinner themselves. They are capable of detecting prey only a few centimetres away from the background. It's no surprise that they are so difficult to catch using mist-nets.

They are also challenging to pick up using microphones, as they can often only be recorded when within one metre of the microphone.[6]

Not every bat species echolocates; Pteropodidae, one of the most diverse families, doesn't use laryngeal echolocation. Several species instead use tongue-clicks or wingbeats to assist them in navigating their surroundings. Phyllostomidae are usually described as 'quiet' bats. They don't produce 130+ db calls to gather information on their environment but, like Pteropodidae, rely on their other senses such as sight, hearing and smell. One notable exception is the Long-legged Bat *Macrophyllum macrophyllum*, the only strictly insectivorous species in the family that can echolocate and shows similar adaptations to fishing bats such as *Noctilio* (bulldog bats).

Studying the echolocation of bats becomes interesting to me when it is used as a non-invasive survey tool. The incredible plasticity presents some significant challenges, but that's what makes it so fun.

Of course, bats are not the only echolocating mammals; cetaceans also use echolocation, as do some bird species. On paper, those two groups could not be more different – one flies, the other one is aquatic; one includes the smallest mammal on Earth, the other the largest; one is incredibly diverse with over 1,400 species, the other only has 70.

Why would two very different animal groups need the same precise sense? The answer: visibility, or lack thereof.[7]

There are very few species of birds capable of echolocating, namely the Oilbird *Steatornis caripensis* and a few species of swiftlets (*Aerodramus* spp.). This ability allows those species to roost deep inside caves, far from most predators.[8]

Notes

1 Elemans, C.P., Mead, A.F., Jakobsen, L. and Ratcliffe, J.M. (2011) Superfast muscles set maximum call rate in echolocating bats. *Science* 333 (6051): 1885–88. https://doi.org/10.1126/science.1207309

2 Fenton, M.B., Jensen, F.H., Kalko, E.K. and Tyack, P.L. (2014) Sonar signals of bats and toothed whales. In A. Surlykke, P. Nachtigall, R. Fay and A. Popper (eds) *Biosonar: Springer Handbook of Auditory Research*, vol. 51, pp. 11–59. New York: Springer. https://doi.org/10.1007/978-1-4614-9146-0_2

3 López-Cuamatzi, I.L., Vega-Gutiérrez, V.H., Cabrera-Campos, I., Ruiz-Sanchez, E., Ayala-Berdon, J. and Saldaña-Vázquez, R.A. (2020) Does body mass restrict call peak frequency in echolocating bats? *Mammal Review* 50 (3): 304–13. https://doi.org/10.1111/mam.12196

4 Wu, H., Jiang, T.L., Müller, R. and Feng, J. (2015) The allometry of echolocation call frequencies in horseshoe bats: nasal capsule and pinna size are the better

predictors than forearm length. *Journal of Zoology* 297 (3): 211–19. https://doi.org/10.1111/jzo.12265

5 Schnitzler, H.U. and Denzinger, A. (2011) Auditory fovea and Doppler shift compensation: adaptations for flutter detection in echolocating bats using CF-FM signals. *Journal of Comparative Physiology A* 197 (5): 541–59. https://doi.org/10.1007/s00359-010-0569-6

6 Schmieder, D.A., Kingston, T., Hashim, R. and Siemers, B.M. (2012) Sensory constraints on prey detection performance in an ensemble of vespertilionid understorey rain forest bats. *Functional Ecology* 26 (5): 1043–53. https://doi.org/10.1111/j.1365-2435.2012.02024.x

7 Fenton, M.B., Jensen, F.H., Kalko, E.K. and Tyack, P.L. (2014) Sonar signals of bats and toothed whales.

8 Brinkløv, S., Fenton, M.B. and Ratcliffe, J.M. (2013) Echolocation in oilbirds and swiftlets. *Frontiers in Physiology* 4: 123. https://doi.org/10.3389/fphys.2013.00123

Subic Bay, Luzon, Philippines, Valentine's Day 2019

I had only one target species for my trip to Subic Bay on Luzon, albeit a big one. The species I was after was the Golden-crowned Flying Fox *Acerodon jubatus*. Depending on who you ask, it's either the largest bat species in the world, or the second-largest after the Large Flying Fox *Pteropus vampyrus*. While they can both exceed a wingspan of one and a half metres (6 ft), from what I understand, *Acerodon* has the absolute record while *P. vampyrus* has a bigger wingspan on average. The reality of these records is that they're often poorly defined and therefore open to interpretation. Subic isn't easy to reach from Manila using public transport, bearing in mind I'd never been to a city this large before and everything was quite confusing. I ended up taking a taxi to save myself the trouble, because I didn't have much time and I was only chasing one species. The taxi driver was also happy to help me ask the locals for directions to the bats.

When we got to the location, the news was grim. The roost had been set on fire a few days prior. My hopes of finding one of the most famous bat species were shattered. I comforted myself with the thought that since I was on a Big Bat Year, any species would count the same as *Acerodon*, even if it wasn't as special. So I asked the driver to find us a bit of forest edge and deployed my bat recorder there. The first species on the recorder, the Wrinkle-lipped Free-tailed Bat *Chaerephon plicatus* (technically *Mops plicatus* now), was a species I would become very familiar with – it's one I would record almost everywhere in Asia. At the time, it was still a very exciting find, though. Alongside *plicatus*, I was also able to record the Javan Pipistrelle *Pipistrellus javanicus* and the Large-eared Horseshoe Bat *Rhinolophus philippinensis*, my first of many Asian *Rhinolophus* species. This family (and genus) has an incredible diversity that's constantly increasing as the taxonomic relationships between the various populations on the many Asian islands are better understood. As night came and I had already recorded quite a few species, we started heading back towards town. Suddenly, I heard a noise – specifically, fruit bat chatter. There are dozens of fruit bat species on Luzon but I wanted to try my best to identify them. It didn't take me very long to spot the small group of four flying foxes high up in the trees. They were barely lit by the streetlights, but were illuminated just enough for me to take record shots. From the back of my camera, I could see that one had a distinct pale crown. It was *jubatus*! I was ecstatic. The other three turned out to be *Pteropus vampyrus*. No matter which bat you believe has the size record, that night, I saw the largest bat species in the world. Species numbers 54 and 55 for the Big Bat Year were incredible ones to see.

Palawan is the Filipino island the closest to Borneo. As a result, it shares more species with Borneo than it does with the rest of the Philippines. It's well known for its Underground River, a cave system that happens to be home to eleven bat species, in addition to being a spectacular geological feature. I visited the Puerto Princesa Underground River twice. The first time, I had no idea how things worked but I had my recorder on the whole time. The identification of Hipposideridae and Rhinolophidae from recordings is often as reliable inside caves as it is outside of them, which is unusual with bats. I felt quite lonely in strictly following the photography ban, as everyone else took photos to their heart's content. The fact that the rule wasn't actually enforced is exactly why I went back a second time, when I made sure to photograph as many bats as possible for identification. I thought the combination of photos and sound recordings would give me the best chance at identifying the many very similar-looking species inside the cave system. After the cruise, I also asked the guards where I could find the peacock-pheasant that I'd missed the first time. Chucky, as he was known, was a very popular attraction – Palawan Peacock-pheasants are very hard to find elsewhere on the island. The sighting felt a bit artificial, though, as the poor creature was so used to seeing people that it just came running up to us.

I paid a little extra for a guide to take me back to my accommodation through the jungle. I was hoping for some decent birding. Admittedly, it was the middle of the day, the worst time for birding, but also my guide couldn't identify most birds. That isn't an issue in itself, as they mainly guide tourists who probably haven't birded before. The price charged was more than reasonable for a one-hour guided walk – but what bothered me was that this guide heavily insisted on taking me birding again the following day for a far less affordable price. He was overly forceful about this, to the point that I ended the guided walk early. Well, we were still headed in the same direction, but I got tired of him asking after the fifth time, so I walked a bit faster to get rid of him. Rather rude perhaps, but there's only so many times I can politely refuse an offer. I'd reached out to Erickson, a bird guide I'd met on Facebook. As he was unavailable to guide me, he recommended a colleague, Will. I was then able to spend the next couple of days birding with a real bird guide. When we met, Will told me that he'd just finished a tour with Birdquest, which had left him exhausted. I was quick to reply that it didn't matter much to me if all we did was relaxed birding – because birding Birdquest-style, meaning every possible bird species had to be searched for and found, is not for me. Despite this early warning, I thought we were going

Entrance to the Puerto Princesa Underground River.

through all the target species rather quickly! We also spent considerable time out in the evening, as that was really the reason I'd hired him. I didn't want to be wandering the jungle on my own. This was also an opportunity for me to get acquainted with the rest of the island's nocturnal life, such as the Philippine Frogmouth and the Palawan Scops-owl as well as non-bat mammals such as the Malayan civet – which we spotted high up in a tree while spotlighting. Perhaps unsurprisingly, the bat list wasn't growing all that fast given that I'd already recorded many species at the Underground River. Combining my trips to the river, the outings with Will and the time I spent walking around the lodge in the evenings, I still managed to score a staggering 13 new species. To this list I was able to add one more right before leaving the island: the Palawan Flying Fox *Acerodon leucotis*, which Will gave me tips on locating. He suggested I spent my last night on the east coast so I could see the bats at dawn, as they head out to offshore islands. I wasn't happy about having to wake up before dawn, but it did give me my 73rd species for the year. The list was growing much faster now. It always did in places of higher diversity. Who would have thought?

Visiting yet another island, I took the ferry from Davao to Samal, where the Monfort Bat Sanctuary is located. It is the largest colony of Geoffroy's Fruit Bats *Rousettus amplexicaudatus*. The cave system is home to 2.3 million individuals. While this would only add a single species to my list, it's one of the must-dos when travelling to the Philippines, at least for anyone interested in wildlife. The ferry itself was a lot jankier than I was accustomed to, but not so much that it actually worried me. It did seem a bit overcrowded, but it was a short crossing – far too short to be concerned about it sinking. I enquired about the cave's location as I was surprised not to see obvious signs for it, given that it was apparently one of the most popular attractions on the island. I asked a couple of taxi drivers on mopeds, and only the second one knew where it was and agreed to take me. It turned out to only be a ten-minute drive away.

On arrival, I was also surprised not to see other people around, again, given what I'd heard of the place. Before my visit, I was really hoping I'd be able to meet the manager of the sanctuary. Unfortunately, as my schedule was tight, I could not go there on any other day, meaning I visited the place on one of the few days the manager wasn't there. It wasn't the first time I'd missed someone because of my schedule, and it certainly wouldn't be the last. That didn't make it any less frustrating, but it's one of the drawbacks of the kind of competition that Big Years represent.

The place wasn't at all what I expected. I expected a cave with a fence to prevent idiots from taking selfies with bats hanging off their ears or something. But no, it wasn't that at all. It's actually a cave system with several sections. I call them sections because, while they're technically part of the same cave, there were several 'pits' where the inside of the caves could be seen from above. Those pits were fenced – for the reasons I just mentioned, I imagine. Each one of those pits gave a different view onto a separate cluster of bats. I don't know if studies have been made into possible social segregation among those clusters. Are the young ones more likely to be on the outside? Or the males? Is it truly random? It would definitely be most interesting to find out, as it's not always possible to study such a large number of bats at such a close distance. The surroundings of the caves were heavily disturbed. Most of what would have likely been forested or shrubby habitats were gone. It was interesting to see the bats thriving despite this loss of habitat. They are likely travelling further to find food than they used to, but that does not appear to cause significant issues.

It didn't take me very long to find the bats, but it took me a while to find the pit that would give me the best photo opportunities. Using a flash was not an option, but that only pushed my creativity, and I spent close to an hour playing with slow shutter speeds in the hope of getting some interesting shots. I eventually got some, but while this wasn't my first dip into long shutter speeds, landscapes tend to be more predictable than bats, making this a definite challenge!

After adding six species on Bohol, I had to figure out where to go for my last night in the Philippines. I wanted to find a place that was readily accessible from

Manila, but that would also allow me to spot some bats. Bram Demeulesteer, a compatriot who lives on Luzon, had told me he'd seen bats on the university campus in Los Banos. As he is primarily a birder, he didn't know what species they were, but he also told me that the location was excellent for birding. It has incredible plant diversity, including many fig trees, which attract all kinds of fruit-eating animals, including bats. It seemed like the perfect place to spend my last night.

The bats, perhaps predictably, were the Wrinkle-lipped Free-tailed Bat *Chaerephon plicatus*, but I got to visit a natural history museum and look for some bats there too, albeit unsuccessfully. This museum started a trend for the BBY, as after that I tried to visit as many natural history museums in the places I visited as possible. I had already come to the realisation that I would have to spend a significant amount of time in cities, and this would be a way to make those stays more entertaining. All in all, this stay was still worth it for me and the surroundings were immensely more appealing than Manila. In about two weeks, I visited six islands, a fraction of the number comprising the Philippines – and as a result, I saw a fraction of the country's bats. However, I think it's fair to say that I'd made a few good choices that helped me get my list from 53 as I left Australia, to 81 as I landed in Jakarta for an entire month in Indonesia.

Peter's Fruit Bat *Cynopterus luzoniensis*.

Short interlude: Diving break, Sulawesi, Indonesia, March 2019

My second passion, next to bats that is, used to be birding, until I became a bit bored of it and turned to scuba-diving for its combination of technology, technical knowledge requirements and fauna. Lembeh Strait, at the north-eastern tip of Sulawesi, is famous for its diving, specifically for its underwater critters. Unlike most diving destinations, the focus isn't on sharks or rays or colourful reefs– it's on small, rare pipefish and incredibly colourful octopuses. This sounded like a perfect destination for me. The diving resort I'd chosen was a posh one, with individual bungalows with their own hot tubs and a shared pool. On the road from Manado, the closest city to the Strait with an airport, to the resort, I was a bit disappointed to see there wasn't much forest left; it had pretty much all been cut down. I was sort of hoping to be able to 'bag' a few bat species while also enjoying time underwater. Unfortunately, the surroundings meant I was unlikely to find much if anything at all. My only hope was that the pool would attract bats for drinking and because fresh water always has insects nearby. Very few bats forage above the sea, so if bats were in the area, chances are that they'd visit the pool.

However, I'd chosen the resort for its diving, not for the bats. This was a short break in my journey, one that would allow me to focus on one of my other hobbies and search for some underwater critters. I don't dive for the sharks, the rays, the stuff that gets posted online all the time. I dive for the tiny animals that you need to spend time and effort finding. Perhaps my love for enigmatic and hard to find animals is why I got into bats in the first place. Large charismatic creatures on a silver platter? Not my cup of tea. The area I'd chosen is known for its Mandarin Gobies *Synchiropus splendidus* (that actually aren't true gobies), the Wonderpus *Wunderpus photogenicus*, a gorgeous species of octopus, the Flamboyant Cuttlefish *Metasepia pfefferi* and the very rare Lembeh Pygmy Seadragon *Kyonemichthys rumengani*. The latter truly is a bit of a Holy Grail. This is a tiny seadragon, coming in at a measly two and a half centimetres. Needless to say, finding it is challenging. It might not be that rare; it just goes undetected on the vast majority of dives.

I only had three nights in Lembeh, enough for two dives a day with one extra night dive. I'd never dived at night in the ocean, so I was excited to try it out. I rapidly realised that I'd be very frustrated diving without a camera. I decided to take an underwater photography course with their resident photographer and rent a camera to solve this issue. It wasn't a high-end piece of kit, but honestly, it didn't matter. I was looking forward to a few days of discovering a new kind of photography I'd never dabbled in before, and doing so with someone I already knew was a very talented photographer, Ben Sarinda. Throughout my dives with him, I discovered that he genuinely enjoyed the camera I was renting, the Olympus TG-5. I was pleased to hear that those who'd worked with it were happy with the device – for a while, that is, until they upgraded to something more high-end. The issue with underwater photography compared to land photography isn't just the price of the camera and lenses; there's also the housing, the port, the strobes or lights, the arms and the floating devices. It adds up quickly, and the financial barrier for entry into this hobby is substantial. But I was happy with the Olympus, and it was great fun to dive with someone who knew what shots worked well with that camera. Ben helped me a lot with the lighting, something particularly tricky to get right

I took a break from looking for flying hairy animals to focus on underwater creatures, such as this sea anemone.

underwater. My previous experience with artificial lighting was all above water, and physics work very differently in these two spheres. During my night dive, Ben and I illuminated the subjects with a very narrow beam for some staggering results. Granted, that meant the shots came almost too easily, but I didn't have much time, and it was a way to make sure I'd leave with images I could be happy with.

At the end of my fifth dive, I hadn't seen the Wonderpus, or the Mandarins, or the Seadragon, but I had seen the Flamboyant Cuttlefish. I thoroughly enjoyed my encounters with more common species such as the Orangutan Crab, which does look a bit like an orangutan, both because of its behaviour and because it's orange and fluffy. My sighting of a couple of pygmy seahorses will likely forever remain in my memory. It was my last dive, and I'd said I would like to try to find a pygmy seahorse as I knew they were in the area, and a couple of fantastic photographs were hanging in the camera room of the resort. Apparently, they're not overly rare, but they are very local. They require a specific kind of gorgonian, the one they've adapted to in order to mimic its looks. We got dropped off by the boat on the way to a dive site the other divers were headed for. The place they dropped us has two of those 'sea fans' that the pygmy seahorses live on. We checked the first one, to no avail. After only a few seconds near the second gorgonian, my guide let me know he'd spotted one. He wrote down on his slate 'There's one here. You have five minutes to find it.' I did my very best to find it, but the five minutes went by very quickly. So my helpful guide wrote that he'd give me an extra three minutes.

After what seemed an eternity compared to the five minutes I'd just spent trying to find the bloody thing, there it was, right under my nose. Their camouflage is truly impressive. It wasn't just the same colour as the gorgonian; it was the same texture too. Usually, I'm pretty good at spotting camouflaged things because the differences in texture lead to differences in how the light bounces, thus revealing their presence. But not in this case. How the guide spotted it in only a couple of seconds is beyond me, but I guess once you've seen them hundreds of times, they can't fool you anymore. Or perhaps they still can – it's the sort of thing you have to spend a few seconds finding again every time you blink. And after spending some time photographing the seahorse, I lost it. The guide had lost it too. A couple of minutes later, we'd find it again. Both of us. In different parts of the gorgonian. This was only possible if there were two of them, and that was indeed the case. I will forever remember my excitement when I first found it. I even laughed – well, as much as is possible underwater with a diving regulator in your mouth. I am really grateful to the guide for not showing the seahorse to me immediately, as that would have made the whole experience far less entertaining. At the end of the day, the search for it was as exciting as the animal itself. That was the sort of experience I wanted to fill my Big Bat Year with, albeit preferably involving bats.

Tangkoko, Indonesia, March 2019

Because I couldn't find bats in the immediate vicinity of the resort, I'd arranged a trip to Tangkoko National Park with Irawan, a local bird guide. This park is known for its macaques, attracting mammal enthusiasts and more generic tourists alike. It's also known for its tarsiers, which tourists are pretty much guaranteed to see on night walks. And then, of course, the park is visited by birders for its special species such as the Lilac-breasted Kingfisher, the Black-billed Kingfisher and the Sulawesi Dwarf Kingfisher, as well as some non-kingfisher birds such as the Minahassa Masked Owl and the Sulawesi Hornbill. We started with a bit of birding that yielded one of the many species of kingfishers present in the area, the Green-backed Kingfisher. It let us get quite good views of it, which was rather lovely as I wasn't used to seeing them so cooperative. They're usually not very fond of humans – or any other mammal, for that matter. Irawan knew this park like the back of his hand and was friendly with all the rangers. As a result, he knew exactly what was happening and where. He didn't know much about bats but like Will, he was keen to show me birds and perhaps some suitable locations for bats; and he was keen to learn more about them, if we found any. He did know of one tree roost. Irawan had seen three bats inside the week before but when we checked, there were 13! They were Lesser Asian False-vampires *Megaderma spasma*, a common bat throughout Southeast Asia belonging to Megadermatidae, the same family as the Ghost Bat I'd seen in Australia. I'd already seen it on Bohol but I had to investigate every lead I was given if I wanted this Big Year to be successful.

The sun went down quickly and we moved to a different area of the park that had a known Spectral Tarsier roost. Like many bats, they roost in hollow trees, and once a tree has been found, they tend to remain faithful to it. Unsurprisingly, as we got there, we weren't alone. The fact that guides and rangers know this roost means that people regularly visit it on night walks and private tours. I feared this would be one of those sacrificial locations of famous species that get publicised to reduce the pressure in other areas. If people want to see tarsiers, chances are they'll target the easy place instead of moving heaven and earth to find one in a remote location. It puts less pressure overall on the species and the

A group of bats is called 'a bunch of oysters'. Here we have Lesser False-vampire Bat *Megaderma spasma*.

habitat. It's not the worst idea in the world, but it does show how much pressure ecotourism can put on certain species.

As it started to get quite dark, patience ran thin for some people there. I was mainly waiting for it to get completely dark so I could find bats. I still wanted to see the tarsiers so waiting a bit longer was not an issue for me. One individual climbed out of the roost tree, jumped a couple of branches, and disappeared from view. I only saw it for a couple of seconds – which was frustrating. Then another one came out, and then a third one. They fought a bit, jumping from branch to branch, and then one of them followed the first one and disappeared too. The third one stayed around long enough for me to get some photos. Probably to reduce the disturbance, tourists on night walks were given dim torches. These didn't seem to disturb the tarsiers at all – but then neither did my brighter torch nor the occasional flash trigger. Maybe this wasn't that sacrificial after all.

Once it got truly dark, we left the tarsiers to do their tarsier things, and I started looking for bats on the recorder. There was surprisingly little activity inside the forest. As we got onto the main trail, though, things picked up, and I was able to record the not-at-all aptly named Nepalese Whiskered Bat *Myotis muricola*, as well as the Black-bearded Tomb Bat *Taphozous melanopogon*. The latter was not a species I'd expected to see this early in my trip, but I knew I was certain to find it in Thailand where it's common. However, getting it on the recorder in Sulawesi was good news because elsewhere, it was sympatric with other *Taphozous* species that have very similar calls, making it impossible to be certain of an identification. While *muricola* is common in Southeast Asia, I was happy to finally find a *Myotis* species that was easy to identify based on its echolocation. *Myotis muricola* calls look significantly more like *Pipistrellus* or *Miniopterus* calls – similar to hockey sticks – than calls produced by other *Myotis* species, which tend to be steeper (with the notable exception of several Neotropical *Myotis* species). Well, to be perfectly accurate here, it wasn't *muricola* but a recently described species that has been split from *muricola* in 2010, *Myotis browni*. I like the name – it sounds tasty.

The bat activity quickly dropped, likely because of the rain, we decided to head back. On the way to the rangers' station, Irawan offered to show me a Sulawesi Black Tarantula. I hadn't seen a tarantula before, so I was definitely keen, but I didn't get my hopes up either because I thought they were wandering predators, not the sort you can reliably spot on any given night. It turned out I was wrong; the tarantula was exactly where Irawan expected it to be, at the entrance of its burrow. It was a gorgeous dark slate-grey with slight bluish hues in it. Undoubtedly far more striking than the name otherwise suggests. Twenty or so metres further down the path, there was another burrow, but no one seemed at home. Its occupant had wandered off but only by half a metre. This individual was much smaller than the one we'd just seen and wasn't as cooperative, as it didn't let me take a single photograph. The first one was extremely calm and allowed me to take as many pictures as I wanted, letting me tweak my settings between each shot to make sure I had something that accurately captured its beauty. I don't think I'd ever been 'wowed' like that by a spider before! The highlight of the night, in my book. These tarantulas are popular in the pet trade because of their gorgeous blue sheen. They're apparently, based on the limited reading I've done, not a particularly easy species though, even by the standards of Old World tarantulas, famous for being more aggressive than their New World counterparts. This is likely a barrier substantial enough to prevent this 'pet' from going mainstream, which is a good thing.

Leaving the lowlands behind, after scoring my thousandth bird species for my life list, the Isabelline Bush-hen, we headed to Mount Mahawu, hoping to find different species of both bats and birds. We made a brief stop at a well-known bat cave. At the entrance, we were greeted by a dreary sight: a mist-net that had been left there, along with a dead swiftlet caught in it. Gua Susuripen was quite a depressing introduction to bat caves in Indonesia. It was also immediately apparent that the bats were unusually shy there. I could hardly find any. All I eventually got was a Sulawesi Horseshoe Bat *Rhinolophus celebensis* that I didn't see – it only flew past my recorder. This species is almost endemic to Sulawesi where it is one of the most common species, though it is also found on Java. It isn't targeted by hunters because it's so tiny. But then neither are swiftlets, yet the one by the entrance had clearly died because of a mist-net. It seems unlikely that only the targeted fruit bats would find their death in those nets. On our way out, my mood was brightened somewhat by the sighting of a Speckled Boobook. Again, it was not a particularly rare species, but I'd grown quite fond of my owl sightings in bat caves, so it was a welcome visit. Well-known bat caves and bats don't go together too well in Indonesia, it seemed!

The famous 'Extreme' Market of Tomohon wasn't somewhere I was particularly looking forward to visiting, but people at the diving resort had mentioned it as being a place where bats are commonly sold. As I'd given myself the challenge of documenting bat-hunting, whenever I could, I added that to the itinerary and Irawan took me there on the way back to the airport for my flight south. The bat trade happening in Tomohon had already been documented quite extensively but photos remain rare. When we got there, I understood why: the whole place is quite gruesome. I even began to feel a bit sick – and normally, I'm not one to be disgusted easily. It's estimated that as many as 500 tons of fruit bats are exported annually to northern Sulawesi from other parts of the island, and from wider Indonesia. It's a much bigger issue than the usual local hunting. The reality is that they've emptied the forests of most of the edible animals and are now having to import them from elsewhere. Interestingly, a recent survey throughout Southeast Asia looking into the drivers of bat-hunting showed that the two most common reasons given in many parts of Indonesia were essentially 'it's a fun practice' and 'it's allowed'. Luckily, mentalities are starting to change thanks to more and more NGOs putting in the necessary efforts to educate people, mainly children. As Irawan explained to me, changing the children's perspective can be a powerful tool to change an entire society's perception. There are ongoing projects to turn the Tomohon market into a green market, a sign that the social norms are shifting in the right direction. Hopefully the wildlife will recover from all this.

Before leaving Sulawesi, there was one more species I wanted to find, one that had been driven to extinction in many northern parts of the island – the Sulawesi Flying Fox *Acerodon celebensis*. Before flying to Seram, I decided to stop by a village close to Makassar in the south of the island. That village is home to a conservation programme where locals have been told about the various bat species present and their importance in the environment. Since then, the locals have been taking tourists to see the bats, in exchange for a small fee. This added income is a significant incentive for them to protect the bats. The colony of *A. celebensis* was huge! Perhaps more surprisingly, the bats weren't very shy at all, which if you've visited Sulawesi, you will know is a rare occurrence. I therefore took my time to take photos and made sure to check that none of them were a different species. The guide who showed me the bats told me that a few of them were Grey Flying Foxes *Pteropus griseus*, but I thought they were, in fact, juvenile *A. celebensis*. This was later confirmed by some researchers who monitor the colony, who told me there weren't any *griseus* left there. I was hoping that *A. celebensis* would be my hundredth species for the year because of the conservation stories that go along with it, but sadly, I was a few species short by the time I left the island. *A. celebensis* was my 92nd bat species and I saw it on the 68th

Sulawesi Flying Fox *Acerodon sulawensis.*

day of the year, meaning that my ration of new species per day, at that point in time, was 1.35. It wasn't quite what I was aiming for, but I had spent some time diving and I'd only just started visiting Indonesian islands. It was about to get a lot better!

During my short trip to Seram, I managed to add another nine species to the list, a little over three per day. Vinno is a guide who was familiar with the Malukus, of which Seram is a part. He had arranged for a few cave visits and helped me ask the locals about bats, I was able to find all my target species. The first bat we found was a Pallas's Tube-nosed Fruit Bat *Nyctimene cephalotes* that flew past us, giving me flashbacks of my sighting of its cousin while in a pool in Tropical North Queensland. Beyond *Nyctimene* and the Greenish Naked-backed Fruit Bat *Dobsonia viridis* (which we also found), my quest focused on the four flying foxes. The locals describe them as the 'Red', the 'Black' (actually two species) and the 'White' flying foxes. This information came in handy because I had been unable to find photos of the Temminck's Flying Fox *Pteropus temminckii*, aka the 'White' one. The prospects of finding a white flying fox were exciting, even if I didn't know what I was looking for. One showed up while we were birding, as Vinno drew my attention to a very pale creature, almost white in fact. It could only be *Pteropus temminckii*, species number 100 for the year. I'd argue it's actually most deserving of that milestone, given its rarity.

The 'white' one, Temminck's Flying Fox *Pteropus temminckii.*

Seram Flying Fox *Pteropus ocularis* (one of the black ones) was the trickiest of the lot; it appears to be more nocturnal than the other three and is more solitary. On the opposite end of the difficulty scale, the 'Red' (Moluccan Flying Fox *Pteropus chrysoproctus*) and the other 'Black' one (Black-bearded Flying Fox *Pteropus melanopogon*) were readily spotted at dusk. For a change, all my targets were fruit bats, but I wasn't going to spend my evenings without turning on a recorder! *Mosia nigrescens*, which I'd already recorded in the Solomon Islands, was extremely common; we even saw groups foraging above the roads. The only new species I was able to add with the recorder was my first ever trident bat, Temminck's Trident *Aselliscus tricuspidatus*. Like many other members of its family, it has a highly distinctive echolocation, making for easy identification even without being able to see the nose.

West Papua, Indonesia, March 2019

West Papua was a part of Indonesia I was very excited to visit. Not only does it have incredible birdlife but it's also largely unexplored as far as other taxonomic groups are concerned, including mammals. I got in touch with Carlos Bocos, a guide from Birdtour Asia who was keen to help me make my short trip there as successful as possible. Unfortunately, for financial reasons, I could only hire him to visit the Arfak mountain range. I also wanted to visit Waigeo, a small island off the bird's head of New Guinea well known for its diving, but I had to do this on my own. Like Seram, it's a paradise for naturalists, particularly birders, with superb endemic bird species. It's also suffering from the same habitat loss issues as most other islands in the world.

Unlike on Seram, however, I hadn't found a paper detailing locations for reliably finding bats, and so I had to figure it out on my own. The owner of the homestay I'd booked didn't speak English and I didn't speak Bahasa Indonesia, making communication problematic. It was difficult for me to ask him where to look for bats or if he knew of bat caves. Luckily, Carlos was happy to talk to the owner over the phone, translating my requests. Once he'd understood I wanted to see bats and other mammals, he became extremely helpful. The fact that I was his only guest may have helped. One afternoon, he came to find me to tell me he'd found a cuscus. Far more similar to possums than to a traditional North African dish, cuscus are marsupials, sometimes diurnal. Understanding what he'd found wasn't hard, because he kept saying 'cuscus'. In fact, as he realised it was the only word I could understand, it's all he would say. He kept pointing to some trees and saying 'cuscus' – yet somehow, it took me forever to see it. The owner probably thought I was blind. After I'd finally spotted it, I even found myself doubting my own eyesight because of how obvious it was! We had a good laugh after that, which is always lovely as laughter is one of the few things that crosses language barriers.

I'd been told that trapping *Mosia nigrescens* was quite the challenge. When I realised they were flying all around my cabin, I decided to put up a net just behind it to try my luck. My endeavour was short-lived as after catching my first bat, I almost caught a careless dog, so I packed up everything as quickly as I could before the dog got entangled and caused a scene. When I opened the bag containing the bat, I found that I'd caught the Broad-eared Horseshoe Bat

Rhinolophus euryotis, an endemic to New Guinea (and its surrounding islands, including Waigeo). It's a species I hadn't recorded yet and not one I had expected to catch in the net. *Rhinolophus* species, especially small ones, are notoriously difficult to capture due to their specialised echolocation. But in this case, it was *Mosia* that turned out to be impossible to catch – they were constantly flying around the net but were able to avoid it very easily.

Arjan Boonman, a Dutch bat researcher now based in Tel Aviv, spent years in Indonesia documenting the echolocation of dozens of species. He told me that there should be a *Myotis* species foraging above the shoreline, most likely the Maluku Myotis *Myotis moluccarum*. It didn't take me very long to record it from the pier. It was interesting to see it above the water as opposed to above the beach. It must have been fishing out at sea, but this behaviour hasn't formally been described yet. I was also hoping to locate some of the fruit bats people had found on the island, but that quest was unsuccessful. The owner even suggested we visit a nearby village where he said there should be bats – but there weren't any. None of the trees had fruit so it was most likely a seasonal occurrence, and I was there at the wrong time of year. While I didn't find most of the known species on the island, I did find one not previously documented as present: the above-mentioned *A. tricuspidatus*. My attempts to catch it were unsuccessful but I did manage to make good recordings of it. Even if I wasn't finding new species for the BBY list, I was making discoveries. And it was about to get a lot better.

Looking for bats in the paradise known as Waigeo.

Waigeo was only my gateway to West Papua. I was ecstatic at the idea of visiting the Arfaks, a remote part of Indonesia, as unexplored as it gets when it comes to bats. The province is visited fairly regularly by birders wanting to see birds-of-paradise, but this region still holds many secrets for mammals too. I'd got in touch with a few guides, and one of them – Carlos Bocos, whom I mentioned above – stood out in particular, as he showed himself willing to compromise on his comfort and expenses to help me make this trip possible. West Papua is expensive, very expensive. For example, renting a car with a driver costs over US$250 per day. Now, that's still half the price of what it usually costs over the border in Papua New Guinea (the main reason I skipped that country entirely), but still, this was well beyond my daily budget, and that was for transport costs alone. Clearly, some creativity was required given the budgetary constraints that I was working with. That creativity came from Carlos.

Carlos and I had done a lot of research and still, we were unsure about what to expect. He'd put together an incredibly promising schedule that involved two nights at each of three different altitudes. Our high-altitude station was a well-known homestay in the Arfaks that gets a lot of visits from birders. As it's one of the few mountains with a road leading up to it, it's easily accessible for those with limited time, such as people on birding trips or myself, on a Big Year! The area surrounding the accommodation looked incredibly wild – as wild as it gets, one could say. It was one of those typical mountain cloud forests, but I love those; the ferns probably add to their appeal. This forest stretched as far as the eye could see, so far that one could easily believe that this ecosystem was untouched, and yet it was far from it. Any stretch of forest with a road leading to it is bound to be regularly visited by hunters. Most of the tree-kangaroos, echidnas, large birds and other mammals have largely been extirpated.

In contrast, away from caves, bats aren't easy to catch and haven't suffered from hunting as much as other mammals in the region. The first bat we encountered on our first evening up there was a Naked-backed Fruit Bat, a *Dobsonia* – a large black bat, likely *magna*, but the taxonomy of the genus is still a bit blurry, especially when it comes to its members in New Guinea. The main issue with travelling to remote regions is that, more often than not, identifications are tentative because of the uncertain taxonomy. However, the biggest reward of visiting remote areas came in the form of another bat.

I'd been recording *Rhinolophus* echolocation calls at around 45–50 kHz, suggesting a relatively large species. I didn't realise this immediately, but there are no large *Rhinolophus* in New Guinea. Then Carlos and I caught a glimpse of the creature, a light-coloured large *Rhinolophus* displaying the genus's typical behaviour of waiting for prey while hanging from a branch on the side of a path. Again, no such thing in New Guinea. Three horseshoe bat species are known on New Guinea: the Broad-eared Horseshoe Bat *R. euryotis*, which I'd seen on

Waigeo and which is much smaller, echolocating in the 52–58 kHz range; the Smaller Horseshoe Bat *Rhinolophus megaphyllus*, which I recorded in Australia (chances are that it is actually a different species) and which isn't known in Papua; and finally, the Large-eared Horseshoe Bat *R. philippinensis* (not to be confused with the Broad-eared Horseshoe Bat…), the largest of the lot but still smaller than our *Rhino*, and also a possible split. Again, it's only known from a few locations in Papua New Guinea, close to Australia and not in West Papua. Echolocation data isn't available for the latter two species in New Guinea, but looking at the nose (from photos that Carlos showed me a few months later), it was clearly different. When we put all these pieces together on the way down to the mid-latitude station, the thought of having found a new species started creeping into our heads. By the time we got back down, after being told we couldn't stay at said station because they'd changed their minds, we'd agreed that it couldn't have been a known species. I'd just fulfilled a lifelong dream! Sure, it was a bit anti-climactic in that it took us two days to come to that realisation. Still, it was quite the achievement. Unfortunately, I couldn't add that species to the Big Bat Year List. However, the Greater Papuan Pipistrelle *Pipistrellus collinus* and the Big-eared Flying Fox *Pteropus macrotis* we got that night were added to it. Those two species are widespread in the area, so finding them there wasn't a surprise.

Cloud forest of the Arfak Mountains, home to some mysterious bat species.

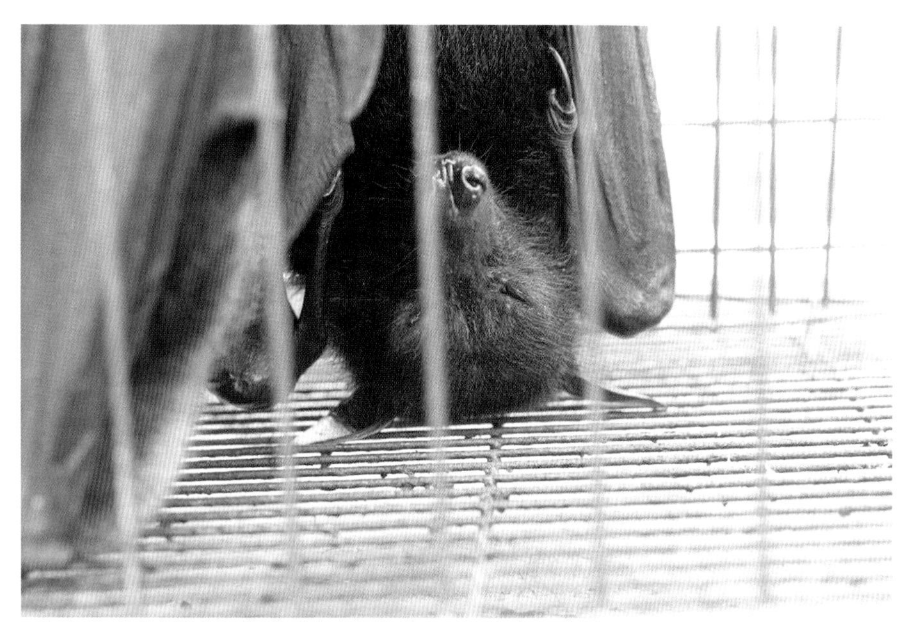

Sadly, not every bat I saw was free to fly – such as this Large Flying Fox *Pteropus vampyrus*.

Down in the lowlands, things weren't as productive, and it wasn't because of the locals; it was because of me. I couldn't wake up at 4 am like Carlos wanted us to, to go batting before birding. The evenings were rainy, very rainy, taking the usual evening outings off the table. After cancelling on Carlos one morning, I realised I had to put more effort in if I was going to make anything of this stay – so the last morning, I woke up at four, and we hopped into the car and headed for one of the few patches of forest that hadn't been turned into a palm oil plantation yet. Bat wise, it wasn't all that great; we got no new species for the list at all. Bird wise, however, now that was a completely different story. It started with amazing views of the rare Papuan Nightjar. Our attempts to take photos were unsuccessful, sadly. Photographs of this newly described species are incredibly rare, so that would have been a real achievement. Other bird highlights included the Twelve-wire Bird-of-Paradise – the highlight of the highlights! Carlos and I saw a lot of bird species during our six days together, far more than I could possibly remember. While I enjoy birding, I definitely do not enjoy it as much as most other birders, and I'm definitely not as happy to wake up as early as them, perhaps to Carlos's disappointment.

Carlos was part of a team monitoring wildlife markets in Surabaya (in East Java). He told me that he regularly saw live bats for sale there and recommended I visit the markets in Jakarta. We'd talked about my other goals for the year, including documenting the bat trade, and this sounded like a good opportunity. I'd hired a driver to do some birding and batting in the evening around Jakarta and he happened to know where the main markets were, which was handy. The first two markets we visited were depressing; they were filled with hundreds of small cages, each containing one colourful bird species I had only dreamed of seeing in the wild. There must have been hundreds of species in those markets, mainly birds but also mammals and reptiles (of the non-avian variety), but no bats. We carried on until we eventually found some. Finding sellers who were open to me taking photos was a challenge though. In the end, as I was starting to feel too depressed to carry on, I got permission, took a few photos of a bunch of Large Flying Foxes *Pteropus vampyrus*, in a cage far too small for them– and then it was time for me to leave Indonesia. Sadly, the last bat I wanted to search for in Indonesia, the Malayan Slit-faced Bat *Nycteris tragata*, turned out to be roosting in a cave that was no longer accessible due to extensive roadworks. What I mean by this understatement is that the road was gone. It had been dug out by machines. Instead, the last bat was one I recorded on a pond in Jakarta, Hasselt's Myotis *Myotis hasseltii*. Thus concluded this epic tour of Indonesia, bringing my list total to 112.

How to become Bat(wo)man – aka a bat conservationist, aka a batter

The world, and the media, are full of misconceptions surrounding science and how one becomes a scientist. While I do have an academic degree, I don't refer to myself as a scientist but rather as a bat ecologist or bat conservationist. However, it's often shorter and easier to just introduce myself as Batman. But how does one become such a hero?

Well, it's quite simple actually! All you need is an interest in bats and a bit of free time. You don't have to have a PhD (or any degree) to be a bat conservationist, but you do need passion, any degree of it. You don't have to be thinking about bats 24 hours a day but it may happen, whether you want it to or not.

The bat conservation community is incredibly diverse. Often, the only thing people will have in common is that they've dedicated their lives, or large parts of them, to bats. The bat conservation community is full of people without academic degrees who make a difference in the lives of both bats and humans by showing an interest in the former and sharing it with the latter. A bat(wo)man's superpower lies in their willingness to share their passion and their knowledge with other people (the more, the better).

Bats can be studied using expensive bat recorders, or cheap ones. They can be studied using expensive thermal cameras or simply with the naked eye. There's so much to learn about them that there are research projects for everyone, no matter how much time or how much money you're able to put into it! You could for example spend a few evenings every summer counting a bat colony when they emerge to see how numbers change over time, as the season progresses. If you're lucky enough to be living where there are flying foxes, why not count them year-round to show how seasons affect their numbers and their schedule. If you prefer doing something during the day, roost searches can be extremely fun: looking for dead trees and hoping to find bats inside.

Not so keen on research? Why not help with outreach or with bat rescue? You could try to find a bat rescue near you and help out when you have time. Or if you fancy working with social media, you could make an impact there – a lot of bat rescuers often dedicate all their time to bats, leaving little to no time to celebrate their work online. The point is, there are dozens if not hundreds of ways you can contribute meaningfully to bat conservation, depending on your interests and skills.

How do you get involved? Facebook, despite all its issues, has made it incredibly easy to get in touch with like-minded people and it's become a great place to get to know your local bat group. If there isn't one, it's also a great place to ask for advice and simply to share your work with bats. Social media has also made it possible for people to organise worldwide events, such as International Bat Night, where bats from around the world are celebrated, along with the people who work with them.

Sepilok, Borneo, April 2019

Leaving the rainy and sometimes nippy forests of West Papua behind, I headed to another rainforest being destroyed by the palm oil industry, this time in Sabah, Borneo. I started in Sepilok with the expected rainforest climate, hot and damp, albeit without the rainforest itself as most of it has been chopped up. Sepilok is the last town before the jungle and it's not entirely fair to say all the rainforest is gone. There are a few remaining reserves and visitor centres. I decided to visit the Rainforest Discovery Centre, a well-known birding site. The Centre also happens to organise night walks, the perfect way to safely be in the rainforest at night with minimal logistical concerns.

My first night walk started with a Giant Flying Squirrel emergence. A couple of nest boxes are visible from the walkway, which is why guides tend to start there. It's a fair way to guarantee a unique sighting for customers. Night walks can be hit and miss, often miss, a bit like whale-watching tours, and these nest boxes help the staff get the satisfaction rates up. I was surprised by the number of participants in that night walk. We got split into three groups, as going on a walk with 40 people is not ideal and the park is large enough to accommodate multiple parties. As we were watching the Giant Flying Squirrels from the walkway, I had my recorder running, hoping that bats would show up, and they did. In fact, I saw the first bat before I heard it. A giant molossid flew over the canopy, drawing the attention of two of the kids. The identification was quick, helped by the acoustics; it was the Naked Bat *Cheiromeles torquatus*. I got really excited, as it was high on my target list! I spent some time explaining to the other participants what the Naked Bat was and why it was called that. A few seemed somewhat interested in bats and kept asking me if I would tick off some new species later that evening. From the walkway, the guide spotted a pit viper through the foliage. True, some pit vipers can be highly faithful to their perches, but I was impressed with the guide's skills nonetheless. It turned out to be a Wagler's Pit Viper *Tropidolaemus wagleri*, a bright green arboreal snake with some stunning patterns. Asian pit vipers are known for their striking looks, and this one did not disappoint. While I'm not a herper by any stretch of the imagination, I do enjoy reptile and amphibian sightings when they are safe, both for the animals and for me.

Further down one of the trails, we stopped because our guide had spotted a bird, a sleeping Rufous-backed Dwarf Kingfisher. This bird was faithful to its little creek, making regular appearances on these night walks, which is worthy of note as the species is otherwise not that easy to see. I wandered ahead to improve my chances of encountering some bats, as there were a few too many lights in the group, and I saw a strange shape on a tree a few metres in front. This weird shape turned out to be a colugo, a Sunda Colugo to be precise. After having seen a colugo on Bohol, I immediately recognised it, but I didn't realise it was a different species. I'd actually seen the entire order at that point, because there are only two species in it. That was my Big Colugo Year wrapped up. Thank for you joining me on that journey. As it's a nocturnal animal, seeing it at night is far more exciting than seeing it sleeping during the day. That's the theory anyway – this particular colugo wasn't that much more active than the one I saw dozing on Bohol. This one might not have been sleeping, but it wasn't moving much either. It seemed a shame to see the best gliding mammal without watching it glide, but even so I couldn't feel too frustrated. This successful night walk ended with the sighting of a Philippine Slow Loris zooming from branch to branch, leaving all the photographers hopelessly trying to get a decent shot of it. I got two slightly blurry

The famous Gomantong Cave.

A rare Orange-haired Wingless Bat.

photos, enough to confirm that it wasn't the recently discovered Kayan Slow Loris, a Borneo endemic that other people have claimed to have spotted in this particular location.

Slow lorises are small primates related to galagos, and to some extent lemurs, more so than they are related to true monkeys. They are one of the few venomous mammals; they produce toxic saliva that kills their prey. So they're not exactly the cuddly kind!

Our last species for the night was a whip scorpion close to the park's entrance. I'd seen quite a few whip spiders already in the various caves I'd visited, and would end up seeing them on all the continents I visited throughout the year – but *Thelyphonida*, as they should be called, are something else. They're not quite as impressive as the whip spiders (*Amblypygi*), but their looks are equally unique! They're also equally harmless, which is reassuring. Another herp highlight that night was the Banded Forest Gecko *Cyrtodactylus consobrinus* that played hide and seek, probably because it didn't enjoy having a light shone at it in the middle of the night.

Unsurprisingly, the reason I was in Sabah was its 'real' rainforest. While degraded habitats, such as Sepilok and its surroundings, can and do harbour good bat diversity, unless I headed for the primary rainforest I knew I'd be missing out on many species. Unfortunately, however, beyond the orangutans, I saw nothing of note thanks to wasting my time and money on a poorly staffed lodge.

After three days of not seeing much, I fled back to Sepilok and got myself booked in for another night walk.

As expected, it started with the Giant Flying Squirrels again but there were no Naked Bats this time. There was very little bat activity and almost no other mammals, in fact. On one of the trails, a large bat, by insectivorous standards, was hanging from a branch. I drew the rest of the group's attention to it, as it was a rather distinctive creature. Beyond the large size, the bright yellow nose was most surprising to see! This is a characteristic feature of Trefoil Horseshoe Bat *Rhinolophus trifoliatus*. Like a lot of other *Rhinolophus* species, it hunts by waiting patiently, perched on the side of a trail, listening for the movement of potential prey.

Evolution

Over the centuries, the position of bats in the evolution of mammals has been subject to a lot of debate. At different times in history, they've been considered as being closely related to rodents, tree-shrews and colugos, and shrews. They're such a unique group of mammals because of their ability to fly, and finding a common ancestor just by looking at them is quite the challenge. With this in mind, it's understandable that the hypothesis that survived the longest was that they were closely related to colugos, the gliding mammals I saw on Palawan and Borneo. This other unique group of mammals isn't capable of active flight but they rely on gliding to move from tree to tree, something bat ancestors also did. However, recent genetic studies have suggested that bats are more closely related to carnivores, ungulates and cetaceans than their somewhat similar-looking but equally weird not-cousins.[1] Those studies concluded that bats belong to the same group as Fereungulata, a grand order which contains ungulates (zebras, rhinos, etc.), pangolins, carnivores and cetaceans.[2]

Unlike the chicken-and-egg conundrum, the question of whether flight or echolocation came first is still under debate. While evidence of flight is pretty straightforward to infer from early fossils, echolocation is significantly more challenging. Bat fossils as old as 50 million years have been discovered, for example *Icaronycteris index*, and these already show the distinctive hand-wing that gives the order its name, '*Chiroptera*'. Despite hundreds of fossils having been dug up from the Messel Pit in Germany alone, very few are complete fossils. Most often, only a few bones are found. In fact, it's estimated that over 70% of bat fossils are missing, meaning we have anything but a comprehensive picture of what came shortly before and after this particular fossil, and others from the same epoch.[3] Full skeletons are even rarer, blurring the lines of the evolutionary history of ancient bats.[4]

Analysis of the fossil records available has revealed that all of them were most likely capable of active flight – that is, they weren't gliders. The features considered by some to be diagnostics of the ability to echolocate are argued by others as being insufficient, leaving no consensus on exactly which ancient bats were capable of it and which were not. The issue with fossils is that they're essentially in two dimensions. This means that distinguishing small bones in the throat or inner ear of a bat is far from trivial. CT scanners have allowed incredible strides in the study of fossils by offering the possibility of creating a 3D model using all the details present in the almost two-dimensional fossil.

The study of the larynx in bat fossils has revealed interesting differences, leading to discussions on what parts are and aren't strictly necessary for echolocation. *Tachypteron franzeni*, a fossil only a little more recent than *Icaronycteris* and *Onychonycteris*, shows incredible similarities with today's members of the Emballonuridae family (sac-winged bats). In fact, it has been assigned as a member of this family. The fact that *Tachypteron*, and *Onychonycteris* and *Icaronycteris* – both assigned to extinct families – lived so close together in time is intriguing.[5]

Icaronycteris index was for a long time the oldest bat fossil ever found, dating back around 52.2 million years. Yet it's commonly accepted that it was capable of both flight and echolocation.[6] *Onychonycteris finneyi* (notice the consistent '*nycteris*', the Greek word for bat, still commonly used in Latin scientific names) from roughly the same era shows no signs of being capable of echolocation, as it lacks the features associated with this ability.[7] However, while those features are found in modern bat species, they may not be essential to echolocate.[8] The question remains unresolved, like many others in bat research.

Notes

1 Upham, N.S., Esselstyn, J.A. and Jetz, W. (2019) Inferring the mammal tree: species-level sets of phylogenies for questions in ecology, evolution, and conservation. *PLoS Biology* 17 (12): e3000494. https://doi.org/10.1371/journal.pbio.3000494

2 Zhou, X., Xu, S., Xu, J., Chen, B., Zhou, K. and Yang, G. (2012) Phylogenomic analysis resolves the interordinal relationships and rapid diversification of the Laurasiatherian mammals. *Systematic Biology* 61 (1): 150. https://doi.org/10.1093/sysbio/syr089. Nery, M.F., González, D.J., Hoffmann, F.G. and Opazo, J.C. (2012) Resolution of the laurasiatherian phylogeny: evidence from genomic data. *Molecular Phylogenetics and Evolution* 64 (3): 685–9. https://doi.org/10.1016/j.ympev.2012.04.012

3 Eiting, T.P. and Gunnell, G.F. (2009) Global completeness of the bat fossil record. *Journal of Mammalian Evolution* 16 (3): 151–73. https://doi.org/10.1007/s10914-009-9118-x

4 Brown, E.E., Cashmore, D.D., Simmons, N.B. and Butler, R.J. (2019) Quantifying the completeness of the bat fossil record. *Palaeontology* 62 (5): 757–76. https://doi.org/10.1111/pala.12426

5 Habersetzer, J., Storch, G. and Schlosser-Sturm, E. (2007, September) Shoulder joints and inner ears of *Tachypteron franzeni*, an emballonurid bat from the middle Eocene of messel. *Journal of Vertebrate Paleontology* 27 (3): 67–104.
6 Simmons, N.B. and Geisler, J.H. (1998) Phylogenetic relationships of *Icaronycteris*, *Archaeonycteris*, *Hassianycteris*, and *Palaeochiropteryx* to extant bat lineages, with comments on the evolution of echolocation and foraging strategies in Microchiroptera. *Bulletin of the AMNH* 235.
7 Veselka, N., McErlain, D.D., Holdsworth, D.W., Eger, J.L., Chhem, R.K., Mason, M.J., Brain, K.L., Faure, P.A. and Fenton, M.B. (2010) A bony connection signals laryngeal echolocation in bats. *Nature* 463 (7283): 939–42. https://doi.org/10.1038/nature08737
8 Simmons, N.B., Seymour, K.L., Habersetzer, J. and Gunnell, G.F. (2008) Primitive early Eocene bat from Wyoming and the evolution of flight and echolocation. *Nature* 451 (7180): 818–21. https://doi.org/10.1038/nature06549

Taman Negara, Malaysia, April 2019

My parents wanted to join me on a part of my trip and I suggested Peninsular Malaysia as a meeting point, since I didn't have much planned there. After four months on my own, the company was more than welcome. I wasn't too sure of where to go – I'd contacted people but hadn't found any leads on things I could do without having to fill in 34 different forms to get a permit. We eventually settled on a few days in Taman Negara National Park, one of the country's most popular continental national parks, followed by a few more days on Penang, an island off the west coast. I'd read quite a few research papers from there, so I figured I would be able to find a few species. More importantly, perhaps, the light schedule meant I'd have plenty of time to relax and spend some quality time with my family.

We were staying right at the very edge of Taman Negara National Park. We were so close to it that I heard Siamangs every morning I was there. The Siamang is a kind of gibbon, a tailless monkey closely related to apes. Unfortunately, I did not get to see any. What I did get to see, from a hide that overlooked a clearing in the forest, with a salt lick to attract mammals, were Sambar Deer and Wild Boar. I only visited the hide early in the evenings, when they were still rather busy. I'd imagine visits later at night would have yielded far more interesting species. That said, I can't say I was expecting to see Wild Boars as I mainly associated them with temperate climates, not with the scorching heat we were enjoying there.

Overall, I spent very little time birding and looking for mammals as I wanted to spend time relaxing and enjoying my time with my parents. Turning on the bat recorder for an hour or so in the evening did get me a couple of new species, such as the Common Thick-thumbed Bat *Glischropus tylopus* and the Greater Asiatic Yellow House Bat *Scotophilus heathii*. I also recorded a molossid which I haven't been able to identify. I thought it might have been the Malayan Free-tailed Bat *Mops* mops. The issue with molossids is that their calls tend to be highly variable. The sequence did not strike me as coming from something like *Chaerephon plicatus*, a prevalent species throughout its range – but an unusual sequence from a common species is far more likely than a typical sequence from an unusual species. Hence, I left it unidentified, like so many other molossid sequences from Southeast Asia.

On an afternoon as hot as an oven, when I was least expecting to find bats, a few came to see me. I was just looking out of our terrace window and saw a bat flying by, and then another one. I decided it was worth investigating. So I headed out of the bungalow and started making my way behind the neighbouring ones. I found nothing at the one right next to ours, but upon close inspection of the one beyond that, I found a bunch of *Cynopterus* hanging under the roof of the terrace. I was thrilled to see them because while they're relatively easy to spot at night, finding a roost isn't anywhere as easy, and it allowed for far better views. They were Forest Short-nosed Fruit Bats *Cynopterus brachyotis*, the most common and the smallest species in the area. Another night of recording bats yielded another few species, including the Narrow-winged Pipistrelle *Pipistrellus stenopterus* and the Lesser Northern Free-tailed Bat *Chaerephon johorensis*. Both are pretty common but by the time we left Taman, my list was at 127 species. I was making great progress!

The drive from Taman Negara to Penang was horrifyingly depressing. I'd seen many palm tree plantations in Sabah, but I wasn't expecting to see absolutely nothing else during a five-hour drive on the motorway. Thousands upon thousands of hectares of forest had been destroyed for this dismal industry. Truly heart-wrenching. When we reached Penang, the bleak landscapes gave way to an urbanised belt around a large patch of forest that protruded in various places. As a result, there was quite a lot of greenery around with various parks and bits of forest poking out. The fact that we would be able to access 'green' areas easily from our flat was one of the reasons I'd picked Penang, as it made it very easy to combine my parents' requirements with my own.

The options for batting were relatively limited because we had to get a car to get anywhere; it wasn't like Taman Negara, where even just the walk to the restaurant could be productive. I convinced my parents to visit the Botanical Gardens in the evening, as I'd seen they didn't close until after sunset. It wasn't my first time in an Asian botanical garden, so I knew I could expect bats. I'd found them to be immensely reliable places to find bats in the early evening. I had about an hour around dusk before the park closed, so I headed off for the wooded areas first – I knew this was where the bats would fly out initially. Indeed, shortly after my arrival, the first bat calls appeared on my recorder, starting with *P. stenopterus* and the Intermediate Long-fingered Bat *Miniopterus medius*. Then came a much larger species, the Greater Himalayan Leaf-nosed Bat *Hipposideros armiger*, which I wasn't able to identify immediately because I hadn't expected to be getting any *Hipposideros* there. It's not a particularly rare species, but it tends to form large colonies, which often requires spacious caves. It's entirely possible that these were just a handful of males roosting inside a dead tree or a small cave. The place definitely did not have any large visible cave; otherwise, I would have visited it. The Small Long-fingered Bat *Miniopterus pusillus* was the second and last new species on the island.

Before going back to my Big Bat Year full-time, I had the opportunity to impress my parents with my wildlife spotting skills – I'd found a Black Giant Squirrel down in the trees behind the car park, from my vantage point on the 17th floor. After careful explanations of its location, they could spot it too. It's not every day that I get to share the sighting of a new species for my list with my parents; a lifer, as the birders would And giant (flying) squirrels are cracking animals anyway.

Kaeng Krachan, Thailand, April 2019

While our intention was to get to the campground of Kaeng Krachan, a large national park right on the border with Myanmar, as soon as we exited the airport in Bangkok, but we realised we'd only get there after the park had closed. Instead, Manu and I had to spend a night right outside the park, at a nature lodge. We made the most of it by spending some time around the lodge with a bat recorder turned on, finding one new species for me: the Malayan Horseshoe Bat *Rhinolophus malayanus*. When we got to the campground inside the park, the wildlife was everywhere. I knew straight away this would be good for bats too. The first bird to greet us was a Black-and-red Broadbill, but we were quickly distracted by the thousands of butterflies flying around dried-up puddles. After grabbing some quick dinner at the nearby restaurant, which was surprisingly cheap given it was most campers' only option, we went on a short walk around the site. Exiting the campground isn't allowed at night for security reasons; encounters with elephants are reasonably likely and could be devastating if they involve distracted or Instagram-addicted tourists. Should the pachyderms decide to visit the camp, there were alert systems in place. Other animals do regularly visit the site – we didn't get to see many, but one such animal was a slow loris, my second species of the year. It wasn't as active as the one I'd seen in Borneo, and we enjoyed good views for a little while before we got distracted by the calls of the Large-tailed and Great-eared Nightjars. I've always found nightjar songs fascinating; they're often quite loud and seem to be coming out of nowhere, well after dusk. In this instance, their calls were intermittently interrupted by thunder. We assumed the storm was headed our way, but it never reached us; it never seemed to have left Myanmar even. A few bats were out and about, but not as many as I'd expected, given the abundance of birds and mammals. The only new species there were the Asian Long-fingered Bat *Miniopterus fuliginosus* and Blyth's Horseshoe Bat *Rhinolophus lepidus*.

We spent another evening on the recorder, as quiet as the previous one, but one new species was added: the Malayan Bamboo Bat *Tylonycteris malayana*.

It must have been feeling right at home given the abundance of bamboo. The whole area looked like secondary forest with its typical rich and dense undergrowth. The *Tylonycteris* genus is fascinating, as they're bamboo specialists. They've got a flat body, allowing them to squeeze into the cracks of bamboo. Their roosts are probably some of the safest out there, at least when it comes to non-human predator protection. The probability of being found by a snake while hiding in bamboo can't be very high! The only drawback of this strategy is that they look silly, being so flat. Before heading back down the following day, we had some more birding time on the other side of the stream, notably with some very familiar birds such as the Yellow-browed Warbler and Grey-headed Woodpecker. Funny how a regular autumn vagrant and a locally extinct bird can make one feel a bit homesick. They were a lot closer to the Belgian birds than things like minivets and broadbills, so I guess it's understandable.

We were up before 5 am, ready to be picked up for the best part of the trip – a five-day guided tourled by P'Kwang, the chair of the Thai bat group, Thaibats. It took us a few hours to reach the 'Painted Bat' village in Eastern Thailand. We were greeted by a group of locals responsible for protecting the bats; this provides valuable additional income when tourists like us come to visit. We had a long chat with our hosts, with the help of a translator, P'Kwang's nephew, because obviously, neither Manu nor I spoke a word of Thai. But we really wanted to learn about the people's relationships with the bats. It was the first time I was able to have an extensive conversation on the topic during the trip. Sure, I'd had a short chat with the village chief in Fiji, but our discussion hadn't detailed how the bats mattered to the locals' lifestyle, or anything like that, because of the language barrier. On this occasion I decided to make the most of the fact we had a translator. While protecting bats and guiding tourists is a significant source of income for the locals, their primary industry is weaving. We got a demonstration, but we also got an example each. Most of the pieces they make take roughly six months, and they are able to sell them for 2,000 baht per piece (roughly US$60), but they also make clothes, pillows and scarves. P'Kwang also spent time explaining the story of this village and the bats.

P'Kwang worked for a Japanese researcher on the Painted Bats, which is how he'd learned about this species and was able to locate them and work with the villagers to help protect them. We had a lot of time to discuss, because they had 'helpers' out in the fields looking for our main target, the Painted Bat *Kerivoula picta*. At first glance, its bright orange and deep black colouration might not look like good camouflage, but it's very effective given its roosting habits inside dead banana leaves. In a region that mainly grows bananas, dead leaves are everywhere. While I would not have minded doing a bit of searching on my own, I have to admit that having experienced and motivated people looking for the bats increased my chances of seeing one by about a million times, give or take.

We eventually got a call telling us they'd found one. We headed towards the spot on a cart, pulled by a two-wheeled tractor attached to a trailer we could sit on. I'd never done that before; it turns out it's quite an interesting way to get around. We were followed for a good portion of the trip by a gorgeous Golden Retriever (or something like that) and some kind of smaller, barky doglet. As we passed through other dogs' territories, there was a lot of barking, occasional teeth-showing, and sometimes even fights. We had a guard dog against other dogs, which I thought was a good idea. Hidden inside one of the leaves of the banana tree in question, there hung A9480, as it was known locally (because of its ring). I asked for a few minutes to take photos in situ before they caught it, as I always prefer photos of roosting bats over those of bats in the hand. After we'd examined it, we let it go, and it immediately became apparent why many people also call them 'butterfly' bats. During the day, one could easily mistake this bat for a butterfly, given its slow and fluttery flight, combined with the bright orange-and-black pattern on the wings. The bat landed in a fresh banana leaf, and I was able to take photos there too – ones I'm still pleased with because the contrast of the orange of the bat and the green of the leaf is just gorgeous. The bat sought shade and so did we. The scorching heat was somewhat suffocating, even though it wasn't as humid as in the rainforest.

Painted Bat habitat.

The infamous Bumblebee Bat *Craseonycteris thonglongyai*, the smallest bat in the world.

Our trip to Kanchanaburi was intense. We visited cave after cave, to the point that they all quickly started blurring into one single cave with lots of chambers and lots of bats. Some were easy to navigate; others required a bit of crawling or sliding down some perilously slippery slopes. The rate at which I was adding species to my list was unprecedented and would remain unbeaten for a large part of the journey. Each cave had three to four new bat species for me to spend time examining, recording and photographing. I'd been doing a fair bit of acoustics work already in the region, so it was nice to get to record some of the bats just to confirm previous identifications. Neither Manu nor I went inside the first cave we visited, as the climb down was particularly challenging, to say nothing of the way back out. Our guide went down and brought back up a bunch of bags; one of them looked heavy and wriggly. It turned out to be a Greater Himalayan Leaf-nosed Bat *Hipposideros armiger*.

For many people, the star of the show during a trip to Thailand is the Bumblebee Bat (a.k.a. Kitti's Hog-nosed Bat) *Craseonycteris thonglongyai*. Sure, I was excited to see it, but I also knew all along that we were guaranteed to see at least one. In fact, we saw Bumblebee Bats in many of the caves we visited, including a female with a pup. Saying the pup was tiny is a gross understatement. We immediately released them without even taking photos, as this sort of stress can lead to a pup letting go of its mum or perhaps even the mum abandoning the pup. But alongside Bumblebee Bats, I was also keen to see what surprises would come our way. And surprises there were! I was raking in new milestones almost every day. The 130th, 135th, 140th, 145th and even the 150th species were all hidden somewhere for us to discover, and we found them all.

The undisputed highlight of this trip was to be found in the last cave we visited. We were in a large cave owned by monks, and we found some bats there; although nothing new for me, the cave itself was gorgeous. As we were about to leave, the monks mentioned a new cave they'd found and built an access path to, so we decided to have a look. It turned out to be a great idea – in that cave was hiding a sizeable group of Stoliczka's Trident Bats *Aselliscus stoliczkanus*, which I recorded and photographed. We then embarked upon capturing one for closer inspection and further recordings. While waiting for P'Kwang to catch one, we all left the chamber to increase the chances as well as to reduce the stress of the operation. I leaned back on the wall of the entrance and felt something was off; I moved a couple of steps to my left and realised I had leaned right against a crevice with a snake in it. It wasn't a deadly species or anything, but I'd argue no snake bite on the back is preferable to a mild snake bite on the back.

Aselliscus, like other trident bats, was on my radar because I thought they were really cool, mainly because of their mysterious trident leaf-nose. But my hopes of actually seeing one had been low. I'd recorded the other species of the genus on Waigeo, but I hadn't seen it. According to the literature, the continental species is relatively rare and confined to a few caves here and there across its arguably wide distribution. Its rarity led me to believe that we wouldn't see the species unless we knew of a specific cave, which we didn't. I had ignored the possibility of finding a new location with some invaluable help from the monks. It was a truly fantastic way of ending this intense yet highly productive five-day tour of Thailand. It was time for us, Manu and me, to head back to Bangkok to catch our respective flights. Unfortunately, we only realised that we were leaving from different airports too late. A visit to the Natural History Museum before my flight the next day allowed me to pretend that I was a bat by putting on a blindfold and ultrasonic proximity sensors. As much as I love bats and as much as I know about them, I make a terrible bat myself!

Reproduction

Most people associate bats with rodents, assuming they produce numerous offspring every year. And at first glance, looking at a bat colony, you might think that's the case. But in many ways, from an ecological point of view, bats are much closer to elephants than they are to rodents. For instance, they live much longer than their weight would otherwise indicate. The record is 41 years for an 8 g species, Brandt's Bat *Myotis brandtii*. Even if the average for that species is closer to 20 years, it's significantly different to the Domestic mouse *Mus musculus*, weighing in at 25 g and living two years on average.

Life expectancy is usually inversely correlated with the number of offspring produced annually. This means that the longer a species lives, the fewer babies it will have in any given year. This makes sense; a short-lived animal only has a short amount of time to raise its youngsters before they have to start reproducing too. By contrast, bats and elephants spend extensive amounts of time carrying their baby in the womb, then raising it, and the babies themselves won't reach sexual maturity straight away. The vast majority of bat species only have one pup a year, although twins are possible, as is the case with humans.

Usually, when only a few pups are born, they tend to be quite big: a bat pup typically weighs about 20% of its mum's weight, some species going up to 40%. It's no wonder that gestation times are long, even in the tropics. In temperate regions, where bats need to hibernate, the egg gets fertilised once spring arrives, suggesting an extremely long pregnancy – but the male's sperm is stored for most of the winter. The gestation length record in bats is found in the tropics, with the Common Vampire Bat *Desmodus rotundus* having been shown to have a gestation period lasting over 200 days. Once the pup is born, it will spend another ten months with its mum before weaning. Again, we're close to elephant territory here.[1] Unlike many other mammals, by the time bat pups are weaned, they have already reached adult size.[2] In fact, during surveys, the only way to reliably tell the age of a bat is to look at the ossification of the finger joints. Depending on the species, those joints will take a few weeks to a couple of years to become fully ossified and therefore opaque. Before that, a light can be shone through to highlight their transparency through the cartilaginous tissue.

In the tropics, some species are polyestrous,[3] meaning they *can* have more than one litter per year, sometimes even exhibiting one shorter and one longer pregnancy, likely to compensate for the differences in food availability.[4] One possible explanation for this is that the mortality levels are higher in the tropics than in temperate regions.[5]

Some species tend to have more than one pup per pregnancy, however, and those also often happen to be migratory species. The Eastern Red Bat *Lasiurus borealis* is a prime example of this – it is regularly known to give birth to two to four pups, making it a true outlier in the bat world. Other migratory species such as the Noctule Bat *Nyctalus noctula* routinely give birth to twins.[5] Because of their migratory behaviour, those species tend to have a shorter life expectancy than similar but resident species. Greater litter sizes compensate for this.

Something quite rare among bats is sexual dimorphism – male and female individuals looking different. It's well known in many bird species, and in some mammals such as lions and many seal (and seal-related) species. But in smaller mammals, it's pretty unusual. In bats, it's found in two families, Phyllostomidae and Pteropodidae, and even there, it's a rare phenomenon. The most obvious examples would be the Visored Bat *Sphaeronycteris toxophyllum*, where the male has a visor, sometimes four times bigger than in females; and the Hammerhead Bat *Hypsignathus monstrosus*, where the male has a huge head compared to the female, hence the name. Interestingly, these two species have also been shown to form long-term harems where one male will live with females all year round.

Notes

1 Greenhall, A.M., Joermann, G. and Schmidt, U. (1983) Desmodus rotundus. *Mammalian Species* 202: 1–6. https://doi.org/10.2307/3503895

2 Barclay, R.M. and Harder, L.D. (2003) Life histories of bats: life in the slow lane. *Bat Ecology* 209: 253.

3 Altringham, J.D. (2011) *Bats: From Evolution to Conservation.* Oxford: Oxford University Press. https://doi.org/10.1093/acprof:osobl/9780199207114.001.0001

4 Kunz, T.H. and Fenton, M.B. (eds) (2005) *Bat Ecology.* Chicago: University of Chicago Press.

5 Knörnschild, M., Von Helversen, O. and Mayer, F. (2007) Twin siblings sound alike: isolation call variation in the noctule bat, *Nyctalus noctula. Animal Behaviour* 74 (4): 1055–63. https://doi.org/10.1016/j.anbehav.2006.12.024

Bengaluru, India, May 2019

Much to my surprise, Meghalaya happened to be a place where I failed to find any bats. There were no urban bats that I could manage to uncover and accessing the wilderness ended up being too complicated to achieve. This made me feel a bit desperate. I even got help from Harpeet Kaur, a local bat researcher, who tried to arrange a visit to a roost of Wroughton's Giant Mastiff Bats *Otomops wroughtoni* but the timing was too tight To compensate for my lack of bats in India thus far, I had to develop alternative plans. One such plan was facilitated by Rohit Chakravarty, another bat researcher whom I'd met through social media. Unfortunately, I didn't have the opportunity to join in Uttarakhand, but who still offered to help me with the Indian leg of my trip in any way he could. He put me in touch with Rajesh Puttaswamaiah, a professor at the University of Bengaluru who works with bats. Despite his busy schedule, Rajesh agreed to take me on a full-day tour where we'd visit two locations where he knew we could find three species I hadn't seen so far. I could tell that he wished he had more time to try for more than that – but honestly, three new species, in the company of a bat expert, was far more than I would have imagined given the number of bad decisions I'd made thus far in the country. After a two-hour drive, a short walk still separated us from the newly built temple we were heading for, carved right into the rock. Rajesh knew this was a good location for Bedomme's Horseshoe Bat *Rhinolophus beddomei*, a close cousin to the Woolly Horseshoe Bat *Rhinolophus luctus*, which I'd seen in Nepal.

As I was busy taking photos of Kiwi, my faithful bat plushie (from the British Bat Conservation Trust) through a window, Rajesh found one. As I focused on it with my camera, I noticed it had a weird shape, and for good reason – it had a pup! I took two photos, and we left. Disturbing a bat with a pup could easily result in the pup's death. After this, we visited a crevice in the rocks nearby, as Rajesh knew it was home to Cantor's Leaf-nosed Bats *Hipposideros galeritus*, a scarce member of the Hipposideridae family. I'd seen a few members in Thailand, but India has a huge diversity within this family; most species I ended up missing for reasons I've already detailed. After a few minutes of looking around and climbing through this rather tricky crevice, we found one bat flying around, never stopping. Getting recordings wasn't an option, mostly because I hadn't brought my recorder with me. Mistakes were made, I'll give you that. The flight looked rather *Hipposideros*-like but providing a species identification based on what we could see was futile – though it was likely *galeritus*. Before leaving the cave, Rajesh insisted on taking a photo of us together featuring Kiwi, the bat plushie that I was carrying everywhere. That made me realise that mascots are genuinely fun things to have. Kiwi was becoming famous.

The other location Rajesh took me to was another temple cave that contained Egyptian Free-tailed Bats *Tadarida aegyptiaca*. While I knew I could tick off that species in Africa, as it's widespread there, I also knew trying to identify it based on its echolocation would be a problematic exercise as it overlaps with several other Molossidae species. So visiting a roost of that species was a good idea. There was also a chance of seeing the Greater Asian False-vampire *Lyroderma lyra* (formerly *Megaderma lyra*), the larger sibling of the Lesser Asian False-vampire *M. spasma* that I'd seen in Thailand and Indonesia. I was pretty sure I'd seen *lyra* in Thailand too, right at the end of our trip, but I wasn't too confident in my identification, so I was keen to have another crack at this species before I met the African members of the family.

The walk to this temple wasn't too long either, and it was nice to be on foot again after another long drive, but it was a steep climb on slippery rocks. As we got there, we checked a crevice near the entry for *Lyroderma* – and there they

Bedomme's Horseshoe Bat *Rhinolophus beddomei*.

A successful attempt at focusing 'blind', Egyptian Free-tailed Bat *Tadarida aegyptiaca*.

were, right where Rajesh thought they would be. They were timid and nervous and wouldn't let me get photos. The *Tadarida* were very chatty, and we could hear them from their crevice, so it didn't take very long for us to find them. Getting good photos seemed like a real challenge here too, both because they were shy and because of the shape of the crevice with lots of bits of rock the bats could hide behind. I immediately realised the bats wouldn't let me focus my lens using a flashlight, so I had to focus 'blind'; I took a couple of shots with flash and checked the back of my camera. Nailed it! It was pin-sharp. I got so lucky, and I was happy because it meant I could end the unnecessary disturbance of the bats there – no need for a long photo shoot. The cave itself was tiny, so we quickly exited and headed back to the car. After a quick snack of mangoes, I spent some time looking at the dragonflies in the area. Dragonflies are a group I am very interested in, in Europe, but I know very little about them further afield, and literature is scarce. As it quickly became apparent that I couldn't look at and take photos of absolutely everything that caught my eye, I'd rarely spent time with them. But they were a nice break from bats. Although I hadn't seen that many bats in India, I was starting to feel tired of this whole journey. The Big Bat Year was a marathon, and I'd been running a 1600 m relay on my own.

Mumbai, India, May 2019

While looking for things to do for the rest of my stay in India, I focused on trying to find people to enjoy any kind of wildlife with, and Raj was the right person for that. When I first got in touch, he told me he didn't know much about bats but that he was more than willing to accompany me to Elephanta Caves and other places too, in search of them. He also started telling me about all the birds we could see and the Leopards in a reserve near the city; I knew I was in the right hands. In between our conversations and my arrival in Mumbai, I'd had a lot more success with the bats thanks to Rajesh, which had elevated my spirits somewhat.

Raj and I met up on Monday morning at my hotel and left from there. While we avoided rush hour on the trains, there were still far more people around than I'm accustomed to. That said, it was nothing like I'd seen in various videos, and I was surprised when local commuters offered me their seats on the train. I definitely would not expect anyone in the West to leave their seat simply because there's a tourist standing. I was still sticking out like a sore thumb, but I felt more like a human than I had in the north, where I was stared at for hours on end. Did they realise this had turned out to be a far more stressful experience than I had anticipated? Was my unease that obvious? I wouldn't be surprised, as this part of the journey was a challenging experience for me. I do not do well in crowded places. While I felt uncomfortable being offered a seat because I couldn't understand why, I thanked them, and I eventually took one so as not to appear too rude. This might sound rather odd to most of you; after all, all that happened was that I was offered a seat on a busy train. However, one of the aspects of my autism spectrum disorder (ASD) that drains me most is the fact that I tend to stress, sometimes panic, when people do unusual things. I'm not sure I would think of giving my seat to someone else, especially a tourist, when on public transport. I'm not sure how to react because I'm not sure why someone would do that – probably just pure kindness, but I see that as a rare gem in our current world. Anything I don't understand, especially when it comes to human relations, tends to freak me out. That is why a journey like this was particularly taxing on my mental health, but it was also an excellent way to perhaps get over this behavioural disorder. While at the time I was unaware I had ASD, all these thoughts went through my head for the rest of the train journey.

As we arrived at the Gateway of India, where we had to take a ferry to the caves, I heard parakeets. At first, I ignored those calls because, well, they're just parakeets; we've got lots of them in Brussels. But then I thought, *Hang on a minute – they're native here! They deserve at least a few seconds of my attention.* Yes, I feel very strongly about invasive species, particularly if they're known to harm local wildlife, as is the case with Ring-necked Parakeets (as shown in a Seville study on bats). From the ferry, the wildlife wasn't all that exciting; it was people mostly. One thing of note, and a rather large one at that, a 284 m highlight I should say, was the INS *Vikramaditya*, an Indian flagship aircraft carrier. I'd never seen an aircraft carrier before, so this was quite the sight, at least for me. While I don't care much for the military, some engineering achievements are worth recognising, and aircraft carriers easily fall into that category.

As far as tourist attractions go, in Mumbai, the Elephanta Caves are definitely very popular, probably because they're a UNESCO World Heritage Site. After passing through lots of stalls selling souvenirs of all sorts, Raj and I started wondering which of the many caves could have bats. Most of them were too small, too well lit and too crowded to constitute a bat roost, so those were already off the table. As our frustration grew bigger because we couldn't find the bats and the crowds felt increasingly overwhelming, I started to lose my temper and took it out on poor Raj, who after all just wanted to help me. We eventually found the bats, above the main statue of Shiva that many people were taking selfies with. Quite a few tourists even took offence that we weren't there for selfies, but we still stood in the way. I was merely trying to get recordings of Schneider's Leaf-nosed Bat *Hipposideros speoris* above the statue. Most people around seemed utterly oblivious to its presence, and I thought it might be a good idea to leave it that way – I didn't know how people would react. I stepped out of the way so people could take their selfies and tried again to get some recordings, unsuccessfully. Leaf-nosed bats often have high-frequency calls that get attenuated very quickly, so they don't carry very far. Had I brought a more sensitive microphone, I could perhaps have recorded them, but what I had simply wasn't cutting it.

We agreed to head out in the evening, looking for Leopards with some stops for bats along the way. The location Raj picked was Aarey Colony. While it's not the Sanjay Gandhi National Park, made famous by *Planet Earth II* with some fantastic thermal imaging footage of a Leopard hunting a pig, it still has a decent Leopard population, leading to regular encounters, often fatal when they involve children. He knew of specific locations where a Leopard was regularly seen, such as one particular bench or the corner of a house. I was impressed that he knew this very precisely – but cats being creatures of habit, I guess that if you spend enough time in their territory, you're bound to see patterns. Unfortunately, while I could see the spots very clearly, it was equally clear that no cats were waiting there for me. So we kept going and started listening to bats; to my surprise, one of the first – after the pipistrelles, of course – was a *Rhinolophus* species, a Rufous Horseshoe Bat *Rhinolophus rouxii*, to be precise. It's a common species, especially there, but that doesn't mean it's necessarily easy to record, given that all horseshoe bats are difficult to detect because of their highly directional echolocation calls. We stopped for a few geckos, too; there were lots of them! Raj also showed me a few spiders that he knew would be around.

There is something really fascinating about looking for critters at night; it's probably the fact that they roam in areas where people wander about during the day that makes this so interesting to me. Their presence makes it look almost like a different world. And this different world includes bats, which of course makes me very happy. While walking and driving around, looking for other potential places to see a Leopard, we stopped in a few spots, listening to bats; I managed to record some more Kelaart's Pipistrelle *Pipistrellus ceylonicus* and the Black-bearded Tomb Bat *Taphozous melanopogon*, nothing out of the ordinary for a semi-rural environment with lots of streetlights. I'd recorded *Taphozous* in a lot of places thus far. While it might become boring to some, I actually enjoyed seeing species like that in many parts of their range. It was as if they were travelling with me. Travelling the world to look for bats can get lonely, OK – don't judge.

At one point, we heard dogs barking and growling. We figured it was just another fight between two dog packs, so we decided to drive past them quickly. As I was on the back of Raj's motorcycle and Raj's friend was leading the way, I was effectively at the rear of our 'pack', which made me extra nervous. Dogs terrify me; I've given up on plenty of opportunities, turned around on paths, changed paths because of dogs, including in my hometown, Brussels. All dogs make me uneasy, some more than others; barking ones terrify me. The ones that are small enough for me to kick into orbit, I can quickly get over, but anything medium-sized or larger tends to make me lose my cool very quickly. Most dogs can sense that, especially a pack of stray dogs. It was enough to trigger a very aggressive reaction on their end, and before we'd realised what was happening, they were chasing us and catching up quickly. I was petrified and started shouting at Raj to drive faster. I'd turned on the back seat of his motorcycle to face the dogs, so as to be able to kick one if it tried to bite me. I've rarely had such a fright in my life; I thought this would end badly. Seeing this, Raj asked his friend to stay a little further back as he knew how to deal with stray dogs. I was just making it worse by panicking and shouting. After what felt like an hour but was probably closer to one minute, the dogs left us, likely thanks to Raj's friend as well as some locals who ran in our direction to help scare the animals off. We stopped for a bit to catch our breath – well, mainly for me. Raj and his friends were really worried about me. I was fortunate to have had such good company that evening. The rest of the night was very uneventful in comparison as we saw no Leopards, nor did we have any further encounters with stray dogs.

As I got back to my hotel, exhausted, I told Raj I would have to skip birding the next day. He'd kindly offered to take me to a place he knew was good for birding. My two main issues with this plan were the early start and the likelihood of stray dogs. I can deal with the first one, occasionally, as I had in the Philippines and Indonesia. Stray dogs, though – I'd had my fill of them already. I know this was one more missed opportunity because of dogs, but it's something I can live with.

Chengdu, China, May 2019

The first few days in China were difficult; I barely left the hotel in the first two days. Instead, I buried myself in some acoustic analysis I'd been contracted for. It did feel rather good to be earning more money than I was spending, for once. These were probably the few days I felt the worst of my entire year. I was about halfway through now, and I had not secured as many species as expected. The following weeks weren't looking very encouraging, as I hadn't managed to get in touch with anyone I could join in the field in China, Taiwan or Japan (except Okinawa). I seriously considered giving up on this Big Bat Year. It had become overwhelming. I'd had a short break in Belgium when I had the opportunity to see friends and family (and bats – I added four species to the list, including the rare Barbastelle Bat *Barbastella barbastellus* and Pond Bat *Myotis dasycneme*). It was fun to meet up with Daan, a Belgian bat worker, to visit the Flemish colony of Barbastelles. I had seen the species in Belgium but in the far south of the country, not in Flanders. I was learning as much about my own country's bats as those in other countries! The break helped to an extent but it wasn't enough. I needed to change things up a bit, spend less time on my own and work on accepting setbacks.

When I eventually headed back out of my hotel room in China, I went back to looking for bats. I'd found a couple of parks in birding trip reports that appeared to be good – for birding obviously, but wherever is good for birding probably is for bats too, to some extent. I'd chosen my hotel based on its proximity to one of the main parks, Qinglonghu Wetland, so I would be able to go there easily without suffering from traffic too much. When I found out I could easily take the underground there, however, everything became far easier. The first time I went, I took a taxi, walked a tiny portion of the park and was somewhat disappointed by what I'd seen. I came back on the tube. The next time, I wanted to spend longer there to find the main birders attractions, such as the parrotbills and Black-throated Tits. The bats there reminded me a lot of the bats I was used to surveying in urban parks in Belgium. Of course, they were different species, but all were somewhat analogous; I found a Pipistrelle, a Serotine, a Noctule. The only one missing would have been a trawling Myotis foraging above one of the lakes.

I wasn't only going to be looking for bats in Chengdu; I wanted to explore a protected area too. The main reason I'd picked Labahe, pronounced 'La-ba-reuh', was for its Red Pandas. I could have picked a number of locations that likely would have good bat species, but knowing this place had excellent birds and pandas made me think it likely had some jewels to be found on the bat front as well. Getting to Labahe on my own as usual was a bit of a challenge, juggling public and private transport to avoid burning a hole in my pocket – but I eventually made it to my hotel, and its gorgeous surroundings full of really cool birds and bats. The forest I could see from my window was a beautiful, typical boreal-type forest in the mountains with lots of moss, dead trees and lichens. The kind of enchanting, almost magical forest I love the most. In many ways, it reminded me of the forests I had fallen in love with while in New Zealand, except the birdlife was very different and richer. Unsurprisingly, so was the batlife! I was able to add another four species, after the four I'd added in Chengdu, including two *Myotis* species. The bats were mostly foraging around the lights along the road and the buildings, but in my quest to find different species, I stumbled across a pond that had great bat activity. I was able to identify the Large Myotis *Myotis chinensis* (I'm telling you, my dislike of common names isn't me being pedantic – they're incredibly unhelpful, as in this case) foraging above the pond. Its calls reminded me a lot of the calls produced by, the Greater and Lesser Mouse-eared Bats *Myotis myotis* and *Myotis blythii*, two closely related species found in Europe.

I was the first visitor of the day on the cable car to the higher parts of the park. I'd been told early was better, which is true for a lot of species. Luckily, that doesn't apply to bats, which is probably why I have stuck to that group after trying out a few other taxa. Early mornings at high altitudes are cold, extremely cold, especially for someone who innocently thought they were going to a subtropical Chinese province, followed by tropical regions of East Asia. I hadn't planned for four degrees Celsius (that's 40°F for our friends across the pond), and the cold was difficult to bear. The sightings of a few exciting bird species helped, but looking for an elusive mammal such as the Red Panda when all you can think of is how cold your hands and feet are getting isn't an ideal experience. China being China, I stumbled upon a building high up in the mountains. This building had a vending machine with its own space heater, as well as toilets. I spent about half an hour with the space heater to recover from the cold before heading back down, without having seen a single 'panda' (Red Pandas aren't really pandas; they're much more closely related to raccoons). I did, however, see a few birds, such as the Sichuan Leaf-warbler, Aberrant Bush-warbler, Rufous-breasted Accentor. They are all somewhat drab birds, to be perfectly honest – but their scruffy look, their resilience to cold, their need to be creative when it comes to finding roosts or perches, are all things that fascinate me.

Sichuan landscape.

Eight new species in China wasn't much, especially given the richness of the areas I'd visited, but I didn't have access to much information regarding roosts and acoustic identification. It's likely that I actually found more species but was simply not able to identify them. My discovery of the pond behind my hotel in Labahe was a great example of my increased determination to plough through what looked like failures and to come up with new ideas to keep adding bats to the list. I was, however, looking forward to my stay in Taiwan, where I'd arranged to meet three different people who'd help me in my quest.

Bats and disease

It's hard to talk about bats without mentioning disease; Ebola, SARS, MERS and SARS-CoV-2 have all been speculated to be associated with bats. Yet there is no evidence of the overspilling of these viruses from bats to humans, as occurs with many other mammal species, for example apes, camels and civets.[1] Viral genetic sequences have been isolated in bats, but no live viruses – proving only that bats have been exposed to the viruses, not that they are capable of infecting other species. This recurring assumption that with every new emerging disease, the virus reservoir is a bat species really hurts people's perception of bats worldwide.

The recent Covid-19 pandemic led to countless killings of bats by people who were scared that they would get infected. However, while a virus close to SARS-CoV-2 has been isolated from a bat species, none of the Covid-19 infections were due to bats; all were due to humans, and yet, a quick Google search shows that a lot of people around the world still link the pandemic with bats.

And this is true for most recent epidemics and pandemics. Suggesting that bats are active reservoirs for all these deadly viruses does nothing to improve bats' fragile reputation. As the early news that Covid-19 may have originated in bats emerged, people worldwide began to kill as many bats as they could find. Those events add to the body of evidence showing that the media's influence on people's perceptions of bats, and perhaps wildlife in general, is substantial.

The reality of it is that while antibodies for Ebola, SARS and MERS have been found in bats, no live virus has been isolated in wild bats. This begs the question of whether they genuinely are the reservoirs that many claim they are, or whether they're just another group of species that can be infected without showing symptoms. Given the intensity of the research on bats and viruses, one would expect live zoonotic viruses to be found in some individuals, but most studies – involving hundreds, sometimes thousands, of sampled bats – have failed to isolate anything more than antibodies.[2]

Obviously, bat virology is a fast-developing field that's still in its infancy, so things tend to change rather quickly and dramatically. But why do we keep going back to bats as the 'usual' suspect when a zoonosis is described? Bats are social animals, making it easier for virologists to sample them for viruses and other pathogens. They are a highly diverse group of species. They often interact with humans, either by roosting in the proximity of humans or by ending up on their plate.[3] However, bats also have a remarkable immune system and none of the bats that showed antibodies were symptomatic. In fact, no bat showing signs of any coronavirus disease or Ebola disease has ever been found!

Bats carry diseases; that's a fact. All animal species do. A recent study has shown that bats do not carry more diseases than any other mammal group when accounting for their diversity.[4] Because bats are the second most diverse mammalian order, right behind rodents, and they often live close to people, including in cities, spillovers are more noticeable.

Most interestingly, there's evidence of social distancing in bats. When bats in a roost are sick, they isolate themselves from the rest of the colony so that other individuals are less likely to be infected. This behaviour has been shown in two families but could well be present in many species. One example of such fascinating behaviour occurs in the Common Vampire Bat *Desmodus rotundus*. *Desmodus* is known for sharing meals with other individuals, including those unrelated to them, but they've also been shown to refrain from doing so with sick individuals,[5] likely to avoid the spread of diseases. The Egyptian Fruit Bat *Rousettus aegyptiacus* has also been shown to exhibit what we humans call 'social distancing' by having sick bats isolate in different parts of the roost.[6]

Too often, the blame is placed on bats to avoid having to accept the real cause; the fact is that our constant ploughing away of the environment puts a lot of stress on wildlife, reducing their resistance to disease.

Notes

1 Calisher, C.H. (2015) Viruses in bats: a historic review. In Lin-Fa Wang and Christopher Cowled (eds) *Bats and Viruses: A New Frontier of Emerging Infectious Diseases*, pp. 23–47. New York: Wiley. https://doi.org/10.1002/9781118818824.ch2

2 Leendertz, S.A.J., Gogarten, J.F., Düx, A., Calvignac-Spencer, S. and Leendertz, F.H. (2016) Assessing the evidence supporting fruit bats as the primary reservoirs for Ebola viruses. *EcoHealth* 13 (1): 18–25. https://doi.org/10.1007/s10393-015-1053-0

3 Wood, J.L., Leach, M., Waldman, L., MacGregor, H., Fooks, A.R., Jones, K.E., Restif, O., Dechmann, D., Hayman, D.T., Baker, K.S. and Peel, A.J. (2012) A framework for the study of zoonotic disease emergence and its drivers: spillover of bat pathogens as a case study. *Philosophical Transactions of the Royal Society B: Biological Sciences* 367 (1604): 2881–92. https://doi.org/10.1098/rstb.2012.0228

4 Mollentze, N. and Streicker, D.G. (2020) Viral zoonotic risk is homogenous among taxonomic orders of mammalian and avian reservoir hosts. *Proceedings of the National Academy of Sciences* 117 (17): 9423–30. https://doi.org/10.1073/pnas.1919176117
5 Moreno, K.R., Weinberg, M., Harten, L., Salinas Ramos, V.B., Herrera M, L.G., Czirják, G.Á. and Yovel, Y. (2021) Sick bats stay home alone: fruit bats practice social distancing when faced with an immunological challenge. *Annals of the New York Academy of Sciences* 1505 (1): 178–90. https://doi.org/10.1111/nyas.14600
6 Stockmaier, S., Stroeymeyt, N., Shattuck, E.C., Hawley, D.M., Meyers, L.A. and Bolnick, D.I. (2021) Infectious diseases and social distancing in nature. *Science* 371 (6533). https://doi.org/10.1126/science.abc8881

Taiwan, June 2019

Joe, a bat researcher based in Taiwan, picked me up from my hotel, and we headed straight to a cave he knew contained the Formosan Lesser Horseshoe Bat *Rhinolophus monoceros*, endemic to Taiwan. The first cave we visited was large and easy to walk into as it had wide, high tunnels. At the end of one of those tunnels, there was a cluster of bats; the tiny *R. monoceros* accounted for the vast majority. However, after reviewing the photographs I had just taken to check the focus, I noticed a different species hiding behind the horseshoes; it was the Chinese Water Myotis *Myotis laniger*. Sometimes photography does make one a better surveyor.

We hopped back into the car and drove a little bit to another area that Joe knew was good for spotting Formosan Woolly Horseshoe Bats *Rhinolophus formosae*, a much larger species of the genus that is also endemic to Taiwan. Unlike *R. monoceros*, *R. formosae* doesn't tend to roost in large clusters or in caves at all, for that matter. They're usually found singly, hanging from roots under overhangs. As we parked, we were greeted by a sign telling people not to leave the paths because there were very dangerous snakes in the area. I reckon that's a much more efficient way of keeping people away from sensitive vegetation than mentioning said plants. As I learned later on, the whole island is covered in these signs. Yes, there are dangerous snakes, but the signs are a slight exaggeration, which is unlikely to improve the reputation of snakes on the island.

There were many small caverns and overhangs a little further down the path and after a quick turn away from it. We started checking all of them, quite methodically. According to Joe, the bats tend to prefer a few of them, but we thought it would be better to check them all. Before long, we'd looked into most of them but without luck. As Joe approached one of the last ones, a large bat flew to the inside of a cavern. I only saw a shadow, but he explained that it had been hanging from a root outside the cavern, as they often do, but had then got scared and flown in. I decided to enter the cavern to try to find it again. The cavern had lots of cave centipedes. They are nocturnal predators that spend the day hiding in caves and caverns. If I can be perfectly honest, they're not the most enticing animals to encounter when entering a cave. I believe they can have a pretty nasty bite, but they rarely bite unless handled. It's only a last resort defence mechanism for them. After investigating a couple of chambers, I found the Woolly Horseshoe Bat again, and guess what? It was hanging from the only root in this small cave system. They do seem to enjoy their roots.

They really like their roots, Formosan Woolly Horseshoe Bat *Rhinolophus formosae.*

Because we'd 'bagged' our two target species but still had time and energy, Joe wanted to visit a tunnel that he thought would offer up some bats. He had visited it a few times before and wanted to see if anything had changed. It took us a while to find the tunnel, as Joe's memories of the place were a bit vague, but we eventually got there, along with a stray dog that decided to follow us. You will have gathered by now that dogs aren't my thing; I wasn't too happy about this. But this dog seemed nice enough, so that I quickly got over my fear in this particular case. It had no intention whatsoever of leaving us alone, anyway. There was one obstacle that stopped us from entering the tunnel, however. Well, I should say two. The small tunnel, too low to stand in, was filled with about 20 cm of water, too much for us to wade in with our walking boots. Neither of us had wellies, and I wasn't too keen on going barefoot into this sketchy-looking water that appeared full of sharp rocks and some suspicious-looking animals. In addition to those suspicious aquatic creatures, there were two gorgeous, bright green pit vipers. They were Chinese Bamboo Pit-vipers *Trimeresurus stejnegeri*, truly magnificent snakes but also venomous. They appeared to be guarding the entrance of the tunnel. It's likely that if we'd tried to enter, they would have considered us as intruding into their territory and might have tried to strike – not something either of us wanted to happen, to be perfectly honest. As neither of us had experience handling snakes this venomous, Joe decided to take a long stick and try to carry it away with it. It took a couple of attempts, but he managed to get the snake to hang onto the branch he'd grabbed nearby. That's when we realised we should have planned this a bit more carefully, as Joe swayed left to right with this highly venomous and agile snake on the end of a stick, sometimes bringing it uncomfortably close to me. Eventually, we decided to leave it on a branch and let it go its merry way, which it rapidly did. It's crazy how easily these snakes can disappear into the vegetation. They stood out when they were just sitting on rocks, but in a shrub they're far harder to spot! It made me wonder how many of those I'd walked past during my time in the Asian tropics.

Given all the trouble this tunnel visit had turned out to be, and the fact there was still one snake guarding the entrance, we decided to call it a day. Joe took me to the train station, to head to Taichung, a city south of Taipei, to meet Keith Barnes, a guide at Tropical Birding. After Joe helped me get a ticket, we asked a passer-by to get a quick photo of us, and I was on my way. On the platform, I was surprised, in the best possible way, to see people queuing in a very organised fashion, following lines marked on the ground. There is no way this system would work anywhere in Europe, or China for that matter. Thanks to the high-speed trains, a very short while later, I was in Keith's car and we were on our way up to Daxueshan, a famous national park with lots of good birds and fantastic landscapes. The gates closed at sunset, but it isn't unusual for bird guides to take clients up there in the evening for nocturnal species, so Keith was confident they'd allow us to drive through.

We spent about 45 minutes birding near the gate and had dinner while doing so. We saw some great species such as the Black-throated Bushtit, the Taiwan Yuhina – a very fancy-looking white-eye family member – a few Laughingthrushes and the Taiwan Yellow Tit, a high-ranking target for a lot of birders visiting the island. I was rapidly getting increasingly confused by the names Keith was throwing at me; as we got to the Steere's Liocichla, I sort of gave up as I'd reached the maximum levels of complexity I was willing to accept from common names. After that, I had to go back to 'Little Brown Bat' level. Or maybe a Taiwan Long-eared Bat would do.

As I struggled with all the bird names, a welcome distraction showed up: a Reeves's Muntjac came down the road. It was fairly shy, and we couldn't get too close to it for photos, but I was excited to be seeing cool mammals again. We drove through the gate at sunset and headed to a road that Keith usually goes along during the day for birding, but we expected that it would also be good at night. We regularly stopped to listen to bats and to search for them with the thermal scope Keith had brought along. At one point, just as I recorded a Mouse-like Serotine *Eptesicus pachyomus*, we managed to find it with the scope as well. Keith made me realise that we were then listening to a processed sound while looking at a processed image of the bat. Technology has really become a powerful tool for wildlife research.

On the way back down, we stopped at a tunnel known to contain bats, but we couldn't find any. I think it was because it was a winter roost. A few species in Taiwan migrate to higher altitudes to hibernate, so I'm assuming that's when those could be found in the tunnel. Looking for bats in said tunnel wasn't all that fun, given that there was the occasional car passing by at high speed.

Back in Taichung, the following day, I had some time to kill before Richard Foster – another Taiwan-based bird guide – was back from a birding excursion in Taipei, so I resumed my series of visits to natural history museums. I was very much enjoying the various approaches and unique perspectives they all offered. When I met up with Richard, he was in the company of a Belgian guy living in Taiwan keen to see some birds. We didn't lose any time; we bought some snacks and headed for the mountains, the same location I'd visited the day before, actually – but we got there much earlier, so we were able to drive all the way to the top. Most birders travel to that park for its pheasants and other rare birds, such as the Taiwan Rosefinch. I don't care much for chicken-like birds, but I do fancy charismatic birds like rosefinches! Continuing our drive up, we stopped for another couple of exciting birds and then we reached the top. It was crowded. The reason for the crowd was the presence of a Mikado Pheasant near the visitor centre. Because we had more time than I'd had the day before, I got to see more of the landscape, yet another cloud forest. I was truly falling in love with these. They're as lush as rainforests but often quieter, and the clouds add an element of mystery. They're also a lot colder than most rainforests. In fact, I was quite

Views of the Daxueshan National Forest.

cold myself – not as much as I had been in Labahe, but still enough to make the birding slightly uncomfortable.

We waited for the sun to set and saw visitors leave the area as the afternoon progressed. Soon enough, there wasn't a single photographer left, and the Mikado Pheasant and the Rosefinch had lost their audience. However, another show was about to start as a group of bats began emerging from the visitor centre's roof. Unfortunately, I was unable to identify the species. On the recorder, however, I was able to identify two others: the East Asian Free-tailed Bat *Tadarida insignis*, one of the few molossid species living this far north in Asia; and the Asian Particoloured Bat *Vespertilio sinensis*, the oriental counterpart of one of my favourite species, the Particoloured Bat *Vespertilio murinus*, which I was hoping to see in Belgium in the autumn as they migrate south. Typically, species like this would be difficult to identify based on their echolocation, but Taiwan being an island has a relatively limited array of species, meaning there are fewer candidates for each possible call sequence.

On the way back down, we found a Taiwan Slug Snake on the road, one of the many snakes on the island. Fortunately, most of the snakes aren't aggressive at all, even when they do have a venomous bite. Being out at night searching for bats and visiting caves meant encounters with snakes weren't all that rare for me. As long as I didn't step on them, I was usually fine. We also spotted quite a lot of mammals, starting with a Taiwanese Red and White Giant Flying Squirrel, a recent split from its continental counterpart that I'd seen in Sichuan. This time it was a lot closer, though, only a few metres away – whereas in Labahe it had been on the other side of the valley. We also found what we'll have to call the Fluffy-tailed Beast for the rest of our days, as we have no idea what the animal was. It was a mammal for sure, and we'll never know any more than that. Even the rat we saw nearby was easier to ID, at least to genus level, as *Niviventer* rats are easily recognisable by their size and white belly.

The mammal show didn't stop there. A couple of Gem-faced Civets playing in the gutter and a Siberian Weasel weaselling through the understorey were definite highlights. I thought I'd already recorded quite a few bat species, but there were two species that I was frustrated not to have found – the Taiwan Long-eared Bat *Plecotus taivanus* and the Darjeeling Barbastelle *Barbastella darjelingensis*. I've always had a soft spot for the plecotine subfamily and knowing that I could find those, and easily identify them based on echolocation alone, was reason enough for me to look. My theory as to why I hadn't been able to record them before is that I was always recording bats after we'd parked somewhere, meaning that we'd already flooded the whole area with the car's headlights. To avoid the issue, I walked ahead of the car, by about 200 m or so, and listened to what I could find. I don't quite know how, but within a few minutes of doing so, I recorded *P. taivanus* and *B. darjelingensis* shortly after. I was over the moon. I'd recorded my two favourite species in this part of the world after what, by all accounts, appeared to be some successful bat troubleshooting. It's worth noting that I didn't come up with this idea all on my own; all plecotine are known to be highly sensitive to light pollution. All I had to do was to apply a bit of ecological knowledge of the species to my actions.

The next day, Richard wanted us to try for a few more birds in the park before birding in a nearby valley. We were able to bag the Taiwan Cupwing and Taiwan Bush Warbler relatively quickly. Both are Little Brown Jobs, as birders like to call them, but I was a lot more excited about those birds than I'd been about the pheasants – their sightings felt very artificial. No matter how colourful, a tame bird is never going to excite me as much as a skulky one hiding in the vegetation that takes a while to emerge. Of course, those can also lead to many very frustrating sightings, which is a downside that can be difficult to accept.

On the way to those birds, we also stopped near a group of photographers who had spotted a Collared Owlet. Daytime owl sightings are always fun, in my opinion, even though doing bat work often means I get a good chance at seeing them at night too. One of the morning's highlights was the Flamecrest, a species similar to the Goldcrests and Firecrests I was familiar with from back home. I like these species, and I was thrilled to see another one. Down in the valley, we found the Taiwan Bamboo-partridge, the Black-necklaced Scimitar-babbler and many others, including the Brown Dipper. While common, I'd have to say the highlight was the latter. Even though I'd already seen this species in Kazakhstan, I have a fascination for dippers and fast stream birds in general.

Okinawa, Japan, June 2019

I had very few contacts in Japan but I'd reached out to Jason Preble, a PhD researcher originally from Hawaii working on two endemic bat species in Okinawa. He was very keen to have me tag along with him on his fieldwork, though he did warn me that the constant downpour would significantly harm our chances of finding bats. Jason picked me up from the airport, and we headed straight for our lodging under pouring rain. We spotted flying foxes from the road. The plan was to drop off my luggage, pick up our kit and head back out, but the weather meant we couldn't go trapping. Instead, we visited a bridge with the two species I was after, the Little Okinawan Horseshoe Bat *Rhinolophus pumilus* and the Southeast Asian Long-fingered Bat *Miniopterus fuscus* – two splits from the Ryukyus of their continental cousins, the Least Horseshoe Bat *Rhinolophus pusillus* and the Eastern Bent-winged Bat *Miniopterus fuliginosus*, respectively. Even if you don't know Latin, you'll quickly realise that the new names simply ended up being slight variations on the 'old' ones... Sometimes it really can be that simple.

The next day, the weather hadn't improved so we went shopping for food while waiting for a break in the downpour to go radiotracking.. It eventually stopped raining for a little while so we took out the receiver and went searching for the tagged bat; we found the spot where the bat was but couldn't see it. We saw the endemic rail and woodpecker. My backpack flooded, as I'd left it open. Not my brightest moment. The forecast said there would be one hour without rain in the evening, so we tried our luck with a harp trap. Because the rain wasn't coming, we opened a mist-net too – but no luck, not a single bat caught. We tried walking around the area, going down to the stream for the endemic *Myotis*, but no luck there either. After missing (or dipping as the birders would say) half the species I could find, it was already time for me to leave the warm and rainy island for a rather cold but dry one, Hokkaido. Hokkaido is at the other end of Japan to Okinawa, which is exactly why I'd chosen those two islands in the first place.

I wish I could tell you all about Hokkaido and its amazing bats, but unfortunately, I did not find a single one during the four days I spent there. Not even one single call on the recorder, despite visiting parks and wetlands and whatnot and the weather being reasonably fair. I cannot fathom this 'failure'. I'd likely been given some erroneous or outdated information at the very least, as the five or so roosting locations I visited were empty and had no signs of recent occupation by bats. I'd been told that there was an industrial complex where bats could be found, so I tried to ask reception if they knew anything about that; they said everything was kept clean, so it was unlikely to be the case, and even if it was, I could not enter. I'm not sure how those roosts were initially discovered, in fact.

The highlight of my trip was the fantastic seafood platter. I treated myself as I knew I had to try their clams and other seafood. A photographer I follow on YouTube, Matt Granger, is also a massive fan of seafood, to the point that he's created a channel where he shares his efforts to find 'the world's best seafood'. While I couldn't afford the sort of places he visits most of the time, I wanted to try at least some of the foods he'd particularly like from Hokkaido. I found a decent-looking seafood restaurant near my hostel and decided to try it. Perhaps, I thought, I should have stayed on Okinawa for a bit longer to get *Murina* and *Myotis*. It was the sort of decision I had to make with very little information available, or sometimes even misleading information – and sometimes, those decisions turned out to be bad ones.

Bats and culture

Bats have been an integral part of many cultures for centuries, if not millennia. The main reason for this, most likely, is that bats, unlike many other mammals, can live close to people. Even today's towns and cities have bats living in them, despite the abundance of light pollution and noise, and the rarity of nature oases. If bats have adapted to modern urban life, it's easy to imagine them living side by side with the Mayas or in medieval castles in Europe. Bats have been depicted by many cultures from all around the world, from the Mayas and their leaf-nosed-bat-adorned pottery to ink paintings from the Far East.

Whereas Christian beliefs identified bats as agents of the Devil, in many oriental cultures, bats were associated with positive outcomes. One example often used to illustrate this is the similarity in Mandarin between the words used for 'bat' and 'luck'.

The Mayas had a 'Month of the Bat', which might indicate that they accepted them as part of their environment, possibly even their homes. Bats were often linked with the underground, which is associated with death and dead people but also with plants. Therefore, the significance of the bat wasn't necessarily as negative as one may think. The Mayas gave their 'God of Death', Camazotz, the face of a bat, a Phyllostomidae. This family is by far the most commonly sculpted by this ancient civilisation. Interestingly, depictions of vampire bats by the Mayas and the Aztecs aren't all that common. This may be because vampire bats only became as prevalent as they are now after the introduction of cattle following invasions by the Europeans.

Bats are oddly rare in Egyptian mythology, despite many deities having been given animal faces. One hieroglyph refers to bats, but their illustrations aren't as common as in other ancient cultures. The goddess 'Bat' has nothing to do with bats. It's somewhat strange to realise this, as the Ancient Egyptians must have been living in relative proximity to many species such as *Rhinopoma* and *Taphozous*, nowadays regularly found in stone buildings.

While many ancient civilisations commonly depicted bats, they also still captivate people in our modern society. The best examples probably are

Dracula, Halloween decorations and Batman. All three offer very different takes on the world of bats. The first one is a fictional character born from the many tales of vampires, most of which initially did not include any bat-inspired features – these only appeared following the stories of the conquistadors who'd come back from the Americas. The book where the character of Dracula first appeared is still considered one of the most popular gothic and horror novels in English literature.

The second is arguably more tailored towards children, as bats have become an integral part of the 'spooky' Halloween symbols, alongside witches, ghosts and pumpkins. One could say that they're probably more closely related to pumpkins, as those do in fact exist.

The third, and perhaps the best known around the world, is Batman, who has been popularised through comics and films for decades. This hero has become so popular that children often dress up as Batman for Halloween, but his image is also used on tuk-tuks in South America and Southeast Asia, and there's even a bus line in Cusco (Peru) with that name. The widespread merchandise based on the DC character has also made its way into the bat worker community, as many of us tend to identify ourselves as Bat(wo)man!

Virelles, Belgium, July 2014

While I've always been interested in nature, I'd argue that my meaningful involvement with nature conservation began when I joined Natagora-Jeunes, a youth group, mainly comprising birders and part of Natagora, the largest French-speaking nature conservation organisation in Belgium. For the longest time, I was living my passion to the fullest, but I was doing so on my own. For the first time in my life, on a cold February weekend in Zeeland, I felt at home in a group, among my peers, young people who were just as passionate about nature as I was. It was a group that took me on as I was, so I had no fear of being an oddball; we all were, to some extent. Perhaps introducing myself as someone who loved doing the dishes was a mistake, as it stayed with me throughout my journey with Natagora-Jeunes, but the revelation showed that I was happy there. Most of our activities revolved around birds, mainly in Belgium, but we had three annual trips abroad: two weekends, one to northern France for the autumn migration, one to Zeeland (the Netherlands) in mid-winter for the thousands of wintering waterbirds; and finally, one longer trip, usually of around ten days, somewhere a bit further afield for some 'exotic' birding. Our flagship activity was a week in the Chimay region (yes, that would be where they make the beer), right by the Virelles lake, an area which we knew like the back of our hand. Early July is a relatively poor time for birding in Belgium – which is why many of us ended up diversifying our activities during this week, from dragonflies to plants and even bats... This is where my story with bats began.

One warm summer evening near Virelles, my friends from Natagora-Jeunes (NJ) and I were invited by Plecotus, the bat workgroup in southern Belgium, for a bat survey to assist with mist-netting. The survey was part of a Life Project dedicated to flowering meadows. Life Projects are funded in part or fully by the European Union. 'Prairies bocagères' was one of the largest ever in Belgium, with a total budget of over four million euros. Like other Life Projects, it had several target species: the Lesser and Greater Horseshoe Bats, Geoffroy's Bat, the Great Crested Newt, the Red-backed Shrike and the Southern Damselfly. Naturally, these acted as umbrella species because they're indicators of high-quality flowering meadows with the presence of clean water. This type of habitat benefits hundreds of species. Umbrella species are usually charismatic species such as birds, pretty butterflies or mammals that the general public can easily relate to. Convincing people to invest in conserving a dung beetle or a mosquito-like insect is rather challenging, and that is where umbrella species come in.

While my first experience with bats was during a trapping session, I discovered acoustics soon after and fell in love with them. While at NJ, I was able to expand my knowledge of taxonomic groups beyond birds, such as butterflies, dragonflies, fish and plants; bats were the ones I immediately became passionate about. I was attracted to the fact that we know so little about them and that they seem to polarise people in a way few other groups do. What I mean by that is that most people out there tend to have an opinion on bats; they either hate them or love them, essentially. The same cannot be said for dragonflies, which seem to leave most people indifferent. Additionally, I was drawn to the fact that bats can be studied and monitored using a wide range of very different methods. Trapping with mist-nets, hibernacula surveys, bat box checks and acoustics are all ways to collect data on bats, and every bat worker seems to have their preference. All these methods allow for different types of data to be collected too, so they complement each other rather well. However, acoustics are rarely a favourite method, probably because they involve the steepest learning curve, and one does not get to see the animals in question, which can be frustrating for many. Yet it's what interested me the most.

Looking for bats in exotic places (there is a real bat in the photo).

I convinced NJ to purchase a bat recorder. After extensive research, I settled on the Wildlife Acoustics EM3+, the first 'affordable' full-spectrum handheld recorder that wasn't without its issues. It took some convincing to get everyone on board as, unlike most other taxonomic groups, the cost of entry into bat work is quite substantial. We agreed that I was to train those interested, which seemed more than a fair deal to me, given it saved me a significant expense. Shortly after I took delivery of the EM3+, I got an opportunity to test it out in the field during our trip to the Camargue in southern France. While bats were only just starting to wake up in Belgium, the milder climate of the Mediterranean meant we had higher chances of seeing and hearing bats, making it the ideal opportunity to trial our newest toy – I mean, tool. I was a complete novice, basing my knowledge on an overly simplistic table I'd taped to the back of the device as a reference. I had no idea what I'd got myself into at the time, and little did I know this would become my passion and profession a short few years later.

The first expedition I organised myself for NJ was a trip to Poland. Finding locations to visit on a budget, that were reachable by road and suitable for birding during the spring public holidays, is quite a challenge. The dates of the spring school holidays vary year on year, depending on the date of Easter. As a result, every other year or so, they are too early for most migrant birds. In Hungary, we ended up being a week or two too early for species such as the Woodchat Shrike, whereas Collared Flycatchers started to arrive as we were leaving. Easter came early in 2016 and so did the holidays. If this trip were going to be a success, we couldn't just look for birds, as many species wouldn't be back yet; we'd have to look for other animals too. We hadn't planned a trip around mammal-watching before, so I thought this would be a good opportunity, and Poland with its wolves, bison (wisents as we should call them in Europe), moose and whatnot was an excellent place for it. It goes without saying that we wanted to find some bats as well. Bird wise, it was pretty clear we'd be far too early for Aquatic Warblers, the main attraction in Eastern European marshes and a species many of us hadn't seen at all, or only briefly in a reed bed on migration in Belgium or in the hands of a ringer. I'd crafted an itinerary that would take us to Bialowieza, the infamous relict primary forest in Europe, and Biebrza, an area of marshland with a high diversity of birds and mammals.

Late March can be a bit of an awkward time for naturalists because wintering birds are mostly gone, but summer migrants haven't arrived yet; hibernating mammals are slowly getting ready for the summer but aren't quite fully active. Bats, for example, have a transition period where they leave their winter roosts but don't regain their breeding colonies just yet, making them somewhat difficult to find. However, this was also the first trip where I sort of knew what I was doing with my recorder, so I could hope to get some lovely bat species that way. Poland doesn't really have any bat species that one can't expect to find in Belgium, but most of them are more common in Poland. The fact that they haven't ruined 95% of their woodlands may be the explanation for this.

On our first evening, we spotted a Grey-headed Woodpecker, a species now most likely extinct in Belgium and that most participants had not seen yet. It turned out to be quite a cold evening, so when the sun went down, I told them that the only bat we could realistically expect in such weather was the Barbastelle. This species is known to forage in temperatures below freezing, looking for micro-moths. It specialises in species that have developed eardrums, as the Barbastelle can 'trick' them by using two alternating faint calls rather than a single louder one as most Vespertilionidae do. What I did not expect, though, was that as soon as I turned on the recorder, a Barbastelle would fly right in front of us! The issue with the model I'd chosen at the time was that it was quite a deaf one, making it less suitable for quiet species such as Barbs – and yet this one must have been close enough to be picked up, and I was thrilled. It was also my first opportunity to hold up my end of the bargain – that is, train the members of our group in bat acoustics. Barbastelles have truly fascinating echolocation, so it was a fantastic opportunity to get started on that. The evening ended on the not-so-distant howling of wolves, which we were all chuffed about as this was one of our target species – one we looked for intensively over the next few days but to no avail.

A little bit later, during that trip, we were able to see some bats and visited a bunker complex on our way to Biebrza National Park. Most of the bats appeared to have already left the bunkers, but a few remained. We were able to find a Barbastelle, which was an excellent opportunity for everyone to see what they look like. Still, there were also better-hidden bats, hiding in crevices, including a Serotine *Eptesicus serotinus* and a Natterer's Bat *Myotis nattereri*. Most of the participants hadn't seen these species before, and this trip likely contributed to bats becoming more of an interest within the group, which I was really pleased about.

Austin, USA, June 2019

The hostel I'd picked in Austin looked nice enough, but more importantly, it was on the Colorado River. I figured the surroundings should be nice enough for me to find some birds and bats without having to drive anywhere. I only had three days in the USA, so my expectations weren't very high; my main goals were meeting Merlin Tuttle and visiting Bracken Cave. I was about to accomplish the former, as I had scheduled a meeting with him the very next day. Since the 1960s, Merlin has been a powerful steering force in bat research and particularly in bat conservation. He is mostly famous for his many staggering photographs of bats from all around the world, used in countless outreach and education programmes; but if you pay close attention, you'll also find many pioneer publications in bat roosting, feeding and reproductive ecology led by Merlin. Since leaving Bat Conservation International, he's been leading his new bat conservation NGO, focusing on education and – something very few other people do – making sure the media get the stories right by reaching out when they don't and offering explanations, along with the occasional photo to improve the visual appeal of their articles. The impact of the media on bat conservation is significant, as it's the main driving force behind people's perception of these creatures. If we just let the media say what they want, it's bound to end badly for the bats. The work Merlin and Merlin Tuttle's Bat Conservation do is therefore crucial to bat welfare. Merlin is an inspiration for a lot of us and his many years of travelling meant we had a lot to share.

Merlin lives on the outskirts of the city, meaning I couldn't walk there and had to get an Uber instead. I would be lying if I said I wasn't nervous when I entered the gated community and tried to find his house. My nervousness grew larger and larger as the house numbers grew nearer to his. I am not a very outgoing person; meeting new people in general isn't necessarily my thing, but meeting people I admire always scares me and makes me ridiculously shy. I don't think I'd be able to count the number of times I have wanted to approach someone to talk to them but couldn't. This was different, though; I knew Merlin was interested in meeting me because he'd accepted my request to visit him. I wasn't worried about that anymore, and yet I was still very nervous when I rang the bell. I was greeted by Teresa Nichta, who works at Merlin Tuttle's Bat Conservation, whom I'd been in touch with as well. My nervousness immediately dropped as it was clear they were as happy to see me as I was to see them. As I sat down with Merlin, we started talking about the various species I'd seen and the circumstances of each, and we went back and forth with stories from the field. At first, it felt a bit unnatural, probably because I was still a bit nervous, and it always takes me a while to get a feel for someone's sense of humour, reactions and non-verbal communication – but soon enough, I was happy to be there and happy to share my tales with the man who's probably got the most field stories out of all of us. I was equally glad to listen to everything he wanted to tell me about. The subject of Bracken Cave inevitably came up, and I mentioned I'd sent an email trying to arrange something with him but had never got a reply. It turned out Merlin hadn't seen it. Despite his busy schedule, he agreed to take me there. He picked up the phone, called someone who worked there and in about a minute or so, he'd sorted it all out – we'd go the next day. I was very excited. After my conversation with Merlin, Teresa showed me the best places to visit around Austin. She recommended lots of things to do, from pools where I could swim (and see Barton Springs Salamanders, an endangered species with strange-looking external gills) to where to eat the best barbecues. I found this wealth of information a bit overwhelming, mainly because I'm not the kind to spend too much time looking for tourist attractions. But I was happy to hear about them nonetheless, and I was determined to try that BBQ place.

While out shopping for a smartphone the following day, I got a text message asking me if I wanted to come by for pizza before heading to Bracken Cave. Of course I wanted pizza. Before that, however, I had some business to attend to. After visiting Tolga Bat Hospital a few months prior, in Australia, I wanted to see what a smaller bat rescue operation looked like. When Lee from the Austin Bat Refuge invited me over, I knew this was the perfect opportunity. I didn't have much planned during the day in Austin, and meeting other people passionate about bats was something that I was keen to do. As I got back from my shopping trip, I got a cab to Lee and Dianne's house, from where they run their operation. The house looked just as you'd expect from people who have dedicated a considerable part of their lives to saving bats. It wasn't very tidy because there were bat cages everywhere; the bats had more space than they did, and the garage had been turned into a bat-processing station for when people brought in bats they'd found grounded in their garden. Lee and Dianne's passion is visible even from the street.

Once inside, I was greeted by a few free-tailed bat pups that Dianne was feeding with a syringe and some formula. Caring for bat pups is tedious because they need to be fed every two to four hours, depending on their age. Every bat carer I've ever met is incredibly dedicated, and I've also always found them to have very caring hearts. Lee and Dianne are perfect examples of this. After meeting the pups, I got to meet the seniors in the house, a couple of old Straw-coloured Fruit Bats *Eidolon helvum*. I hadn't seen those yet, but I was sure I'd spot them a few weeks later, so I wasn't frustrated I couldn't count these as they weren't wild! They couldn't fly anymore; they had mostly turned bald, and they were having a pleasant and comfortable end of life, lying down in a very stylish hammock Lee and Dianne had made for them. They'd come to the refuge after a life of being educational bats at a local zoo. Once the zoo couldn't care for them anymore, Austin Bat Refuge took them in, and they've been living in Lee and Dianne's living room ever since.

Lee was very keen to show me their outdoor flight cage, where they keep most of their bats in care. Again, it was evident that bat care was an integral part of their lives because, well, there was no garden anymore. The garden had become the flight cage. It was a good ten by six metres and probably close to three metres high. There was even a little pond in the middle where the bats could drink, but that also served as a food source because of all the insects that inevitably turn up in any volume of water that's left alone for longer than three and a half minutes. There started the frustrating part of the tour: a lot of the bats they had in care were species I hadn't seen yet, and would be rather tricky to find around Austin. I hoped I would get some decent diversity at Bracken Cave, not just millions of Mexican Free-tailed Bats *Tadarida brasiliensis*. The species they get the most at the refuge is the Eastern Red Bat *Lasiurus borealis*. Often, they're found grounded after people have trimmed down the dead leaves of palm trees. Their reddish colour allows them to blend in quite nicely in there, not unlike the Painted Bat *Kerivoula picta* I'd seen just two months prior. Most of the red bats that come into care are pups, often orphans as the mum flew away to escape the deadly trimming. It's one of the few species of bats on earth that's known to commonly have more than one pup, between two and four.

Near the entrance of the flight pen, there were a few small cages that each held a few red bat pups, all bundled up together, hanging from the ceiling. They were adorable! They varied quite drastically in age; some still needed formula, others were a lot closer to being able to fly. They all seemed equally happy to be involved in cuddle puddles, though, as they would do if their mum were present. One or two cages had a mum with pups, both hers and adoptive ones, too, when needed. Bat pups must learn how to be bats when it comes to flying and good use of echolocation, so a mother's care is difficult to replace in captivity. Still, Lee and Dianne have good release success rates with their pups, likely because they can learn to fly and find food in a safe environment. It is effectively a form of soft release like they do with 'crèches' in Australia, except that instead of being given fruit, they're provided with a pond that 'produces' food.

There weren't just red bats at the Refuge, however. Lee also showed me a single Cave Myotis *Myotis velifer*. Unlike many other North American *Myotis* such as the Little Brown Bat *Myotis lucifugus*, it has stable population trends. The Little

Brown Bat went from being one of the most common bats, with populations numbering millions of individuals, to an endangered species, possibly down to a few hundred thousand individuals. The culprit? White-nose syndrome, a deadly disease caused by an invasive fungus that swept through North America in record time. There was a bat box at the other end of the pen and a bunch of palm tree leaves. The bat boxes contained 'Pipistrelles', as they call them over there. They're not from the genus *Pipistrellus*, but they look alike and have a similar ecology; they're more correctly called Evening Bats, but anyone who argues about which common name is best should just use scientific names. Although, in this particular case, the scientific name is *Nycticeius humeralis*, which doesn't exactly roll off the tongue.

The palm leaves were there for Western Yellow Bats *Dasypterus xanthinus*, a rare species this far east but one I wanted to see out in the wild too. I wanted to get some recordings at the refuge to be able to compare them against the ones I'd collect from free-flying bats, but I hadn't brought my recorder with me. I had also run out of time and had to head to Merlin's already. When I arrived, Teresa greeted me again, but I also met Merlin's wife, Paula. We sat down in the living room for a little bit while we waited for Merlin to arrive. I spotted my first ever hummingbird in the garden, as they'd put up a feeder for them. It was a female Ruby-throated Hummingbird, a ubiquitous species, but I didn't care; I was still really excited by it. Paula must have thought I was a bit crazy to be

Bracken Cave and its Mexican Free-tailed Bats *Tadarida brasiliensis*.

so enthusiastic about such a common bird, but that's what happens when you spend 23 years of your life on a continent that doesn't have hummingbirds. As I got back into Merlin's office, a copy of his latest book was waiting for me. He'd signed it ahead of time with 'Thank you for everything you do for the bats', which coming from him was incredibly meaningful.

We left right after dinner for Bracken, because Merlin wanted to ensure we'd have enough time to set everything up before the emergence. We'd based our ideal ETA off the information we'd got from the emergence the day before, and we tried to get there a bit earlier, just in case. And it turned out to be a good idea. We took a few minutes to set up Merlin's camera and flashes, and then the bats were already flying out, almost an hour earlier than the day before. Luckily, Bracken has a lot of them; that's what it's notorious for, after all. I was witnessing one of the greatest wildlife shows on earth and this time, it wasn't on a small screen, narrated by David Attenborough. It was all very real and I was witnessing this show in the presence of one of the world's most prominent bat conservationists, no less. Millions of bats flying above our heads, mostly not bumping into each other – don't believe people who tell you bats *never* bump into each other, that's a lie. The show was so impressive that even the bench I was sitting on fell over, while my camera and I were still on it. I was unscathed but the camera did suffer some damage to its screens.

Wind turbines

Climate change is considered one of the gravest threats to biodiversity,[1] and alternatives to fossil fuels are widely sought to curb its effects. Solar power, hydroelectricity, geothermal energy all have their benefits and issues, but wind turbines are particularly controversial for birds and bats. Wind turbines have probably always been controversial, either for their impact on the landscape or on wildlife. Historically, the focus has been on birds, as the presence of dead raptors and whatnot were always an obvious sign that something was wrong. A dead 20 g bat is a lot harder to find. Early research showed that bats were much more at risk from wind turbines than birds; they simply were not found as often.[2]

The threats themselves are somewhat different; for birds, it's collisions and detours when they try to avoid the turbines. For bats, collisions are a factor, but there's the added risk of the turbines themselves being attractive to bats, possibly because of the insects that are themselves drawn to its heat. Current research tends to indicate that bats are attracted to wind turbines, thus exposing them to a greater risk than birds, which instinctively avoid them.[3] While the probability of getting hit by a blade is relatively low, many deaths are believed to occur following barotraumas – injury caused by a change in air pressure.[4] The lower-pressure pockets created by the turbulence around the blades can cause the bats' lungs to explode.

Signs of bat population declines because of the growth in wind energy have already been described.[5] Perhaps even more alarming than the scale of the damage done by onshore wind turbines is the potential impact of the even faster-growing offshore wind energy industry. The behaviour of migrating (or even foraging) bats out at sea is still very much unknown, and offshore wind parks are growing like mushrooms in the North Sea for example. Early evidence suggests that they will have a devastating impact on populations of migrating bats and birds.[6]

Much of the current research focuses on reducing the impact of wind turbines on birds. A paper from Scandinavia showed that painting one of the blades black reduced bird collisions by as much as 80%.[7] The same study accidentally discovered that Ptarmigan mortality was not due to hitting the blades but the shaft. A follow-up study showed that painting the shaft black would significantly reduce Ptarmigan mortality.[8] The issue of mitigation with bats is trickier because they are attracted to the turbines or cannot avoid them. The solutions are, therefore, a lot more complex and could be divided into two main categories: active and passive.

Passive solutions include turning off the turbines in low wind or when bat activity is high, for example during the swarming or migration periods.[9] The data upon which those schedules are based has to be collected beforehand, and often, there is no follow-up to check on the effectiveness. Even when there is a follow-up monitoring, it is limited to a recorder mounted on the nacelle, which is an insufficient method of assessing bat activity around turbines because it is impossible to record the smaller species at a distance beyond 50 m, which is an issue when many of the blades currently in use exceed 60 m in length. Yet simply by not spinning the blades at low wind speeds, more than 80% of bat mortalities could be prevented – but this easily implemented knowledge remains mostly unused. Active solutions are still very much under development but include sonar-jamming speakers. Acoustic jamming is limited by the same physics rules as the recorders on the nacelles; air attenuation leads to ineffectiveness towards the edge of larger turbines.[10]

It is essential to continue documenting the impact of wind turbines on these animals.

Notes

1 Garcia, R.A., Cabeza, M., Rahbek, C. and Araújo, M.B. (2014) Multiple dimensions of climate change and their implications for biodiversity. *Science* 344 (6183): 1247579. https://doi.org/10.1126/science.1247579
2 Voigt, C.C., Lehnert, L.S., Petersons, G., Adorf, F. and Bach, L. (2015) Wildlife and renewable energy: German politics cross migratory bats. *European Journal of Wildlife Research* 61 (2): 213–19. https://doi.org/10.1007/s10344-015-0903-y
3 Richardson, S.M., Lintott, P.R., Hosken, D.J., Economou, T. and Mathews, F. (2021) Peaks in bat activity at turbines and the implications for mitigating the impact of wind energy developments on bats. *Scientific Reports* 11 (1): 1–6. https://doi.org/10.1038/s41598-021-82014-9
4 Baerwald, E.F., D'Amours, G.H., Klug, B.J. and Barclay, R.M. (2008) Barotrauma is a significant cause of bat fatalities at wind turbines. *Current Biology* 18 (16): R695–6. https://doi.org/10.1016/j.cub.2008.06.029
5 Frick, W.F., Baerwald, E.F., Pollock, J.F., Barclay, R.M., Szymanski, J.A., Weller, T.J., Russell, A.L., Loeb, S.C., Medellin, R.A. and McGuire, L.P. (2017) Fatalities

at wind turbines may threaten population viability of a migratory bat. *Biological Conservation* 209: 172–7. https://doi.org/10.1016/j.biocon.2017.02.023

6 Solick, D.I. and Newman, C.M. (2021) Oceanic records of North American bats and implications for offshore wind energy development in the United States. *Ecology and Evolution* 11 (21): 14433–47. https://doi.org/10.1002/ece3.8175

7 May, R., Nygård, T., Falkdalen, U., Åström, J., Hamre, Ø. and Stokke, B.G. (2020) Paint it black: Efficacy of increased wind turbine rotor blade visibility to reduce avian fatalities. *Ecology and Evolution* 10 (16): 8927–35. https://doi.org/10.1002/ece3.6592

8 Stokke, B.G., Nygård, T., Falkdalen, U., Pedersen, H.C. and May, R. (2020) Effect of tower base painting on willow ptarmigan collision rates with wind turbines. *Ecology and Evolution* 10 (12): 5670–9. https://doi.org/10.1002/ece3.6307

9 Măntoiu, D.Ş., Kravchenko, K., Lehnert, L.S., Vlaschenko, A., Moldovan, O.T., Mirea, I.C., Stanciu, R.C., Zaharia, R., Popescu-Mirceni, R., Nistorescu, M.C. and Voigt, C.C. (2020) Wildlife and infrastructure: impact of wind turbines on bats in the Black Sea coast region. *European Journal of Wildlife Research* 66 (3): 1–13. https://doi.org/10.1007/s10344-020-01378-x

10 Voigt, C.C., Russo, D., Runkel, V. and Goerlitz, H.R. (2021) Limitations of acoustic monitoring at wind turbines to evaluate fatality risk of bats. *Mammal Review* 51 (4): 559–70. https://doi.org/10.1111/mam.12248

Sinaloa, Mexico, June 2019

Lorenzo and I did the same Master by Research at Imperial College London, and we'd grown close when I was having a bit of a rough time. I was excited to see him again. I flew from Mexico City to Mazatlán, where Medardo, a colleague of his, picked me up as Lorenzo was in the field. We quickly found each other, as it was a small airport with very few people, and headed straight for La Cruz, a small town that happened to be the headquarters of the Snowy Plover research team – that is, Lorenzo and Luke, his supervisor. The state of Sinaloa is well known as home to one of the most famous and deadliest cartels in Mexico. Most of its sinister fame was acquired in the 1980s and 1990s, when Joaquín 'El Chapo' Guzmán was leading it. Sinaloa is a rather large state, and while not every part of it is subject to violence, most of it is still controlled by the cartel. Because it's rather touristic, Mazatlán is safer than most of its other cities and towns. Culiacan is often featured in newspapers because of the somewhat regular shoot-outs between cartel and the so-called *federales* that occur there. About halfway between Culiacan and Mazatlán is La Cruz. We were definitely well inside cartel country, but because the team had taken the necessary precautions, it was pretty safe for us and we would not be perceived as a threat. While I trusted Lorenzo and Luke's view on the matter, I knew that my parents might not. As a result, I avoided being too specific in naming the locations I was visiting, in case the names 'Sinaloa', 'Guadalajara' or 'Jalisco' triggered unwanted anxiety.

My arrival coincided with Lorenzo and Luke's first fieldwork break in months. Perfect timing, I thought. I met them in La Cruz and soon after, we were on the way to a nature reserve near Cosalá. The plan was to do some bird-ringing with some UNAM (Universidad Nacional Autónoma de México) students, but we had arranged some bat-trapping too with the help of a local bat expert. Despite this being a break for Lorenzo and Luke, we had a lot planned. Bat-trapping got us only one bat but a new one for me, Waterhouse's Leaf-nosed Bat *Macrotus waterhousii*. The fact that it wasn't yet on my list wasn't too surprising, given that it was my first time trapping in the New World. However, not only was the species new to me, but it was also my first encounter with the very diverse Phyllostomidae

Landscape in Sinaloa.

family. Little did I know then that I would end up hating them only a few months later. At the time of writing, it's still my least favourite family, as some can bite rather painfully when being untangled from a net – nothing like the 'gentle balls of fluff from the Old World', although others might disagree with this statement. Am I biased? Absolutely! But we all have our favourites, don't we? By no means does this stop me from conducting research in the New World, though. I think perhaps what makes it harder for me is their sheer abundance. Trapping in the Old World is rarely as overwhelming.

Jalisco, Mexico, June 2019

I had found a paper offering detailed information on the locations of a few bat caves in the state of Jalisco, which is why I decided to go there next. I hadn't visited any bat caves in Sinaloa, so I still had a great many species to find. The climate and species diversity also differ significantly from what I hoped to find in Yucatán and Campeche, so it felt like the perfect place to explore. However, as I didn't know anyone in Jalisco, the logistics would be challenging. I hired a bird guide, Jesper, who agreed to help me find those bat caves and also do some birding along the way. Hiring a guide also brought some comfort as many of the place names I'd see there would be too familiar for me; I struggled not to think about the threat posed by the cartels. After a long bus ride, and travelling to an entirely different time zone, I arrived in Ciudad Guzmán where I would meet Jesper.

We left early the following morning to go birding in the hills. Our plan was to be there before sunrise to look for owls, nightjars and other nocturnal creatures. Specifically, our targets were the Great Horned Owl, Mexican Whip-poor-will and Buff-collared Nightjar. We managed good views of the first two but missed the third, or 'dipped it' as birders say. We saw a lot of birds that, mostly, I've forgotten about. That's one of the reasons I don't use bird guides more regularly; I'm not after a big bird list. Obviously, during my world tour, I was focusing on bats; birds were welcome extras, but I couldn't do much research on the bird species I could find because all my time went on bats (though sometimes on some other cool mammals too). What often happened was that the guide would give me a bird name I'd probably not heard before, I'd find the bird – most of the time anyway – and then we'd move on to the next. That's not my kind of birding. However, one bird I do remember seeing was the Red Warbler, an unrealistically bright red, well, warbler. Definitely one of the prettiest birds of my entire journey.

The vegetation on the hills in Jalisco consisted mainly of pine trees, which I hadn't been expecting to find there. The higher altitude meant the climate was more suitable to conifers, but pines aren't exactly the first plant one thinks of when thinking of Mexico. That said, the habitat looked ideal for bats; at least, it would have looked good in temperate regions. I had no opportunity to check because we visited a nearby lake, Laguna Zapotlán, instead. My goal for that lake was to find a Yuma Myotis *Myotis yumanensis*. I'd been told it should be common on water bodies in the region, which turned out to be true as I only had to wait a couple of minutes before one showed up, and I was able to get recordings.

Beyond birds, the reason I'd hired Jesper in the first place was to have some assistance in locating the caves I'd read about in a published study. While he didn't know the site specifically, his knowledge of the area (and his Spanish) would be valuable assets. As we got to the location of the first cave that I wanted to visit, we asked for more precise directions to where the cave might be. Nobody seemed to know, so we started looking around. I quickly found a very small cavity with one bat inside, the Grey Sac-winged Bat *Balantiopteryx plicata*, a small emballonurid species. We spent a considerable amount of time scouring the hill in the scorching sun, hoping to locate the bat cave. It quickly became a painful task to pursue, and we eventually gave up. We were briefly greeted by a snake crossing the trail in front of us on the way down, too quick for any chance at identifying it, unfortunately. While looking around for some birds, I saw three Northern Jacanas. I was surprised to see them there because I was unaware they inhabited the New World. It goes to show how little preparation I had time for when it came to birds.

That's when Jesper pointed out a Tiger Heron high up in a tree above our heads. Herons high up in trees will always look weird to me. After spending some more time talking with the locals, they invited us to share lunch with them. Understandably, given I spoke no Spanish at all, the lunch was quiet on my end, but Jesper and the family seemed to exchange valuable information. They told us about a place on the other side of the hill that should have cavities, though not actually a cave. Once this lovely lunch was over, we were on our way there. The area contained a collection of ponds with some rock formations behind them. We were greeted by a rather aggressive group of geese and a far cuter and far less aggressive Least Grebe. Finding the cave did not take much time – it quickly became apparent that it had to be within the rock formation on the other side of the ponds.

After some very light climbing, we found the opening of the cavity, and it was indeed tiny, barely a square metre, and it wasn't deep either. I'd say the entrance chamber was a little over a metre in depth, and there was a small tunnel going to, presumably, a larger chamber further within the rock formation. None of it was accessible for a regular-sized human, so I had to settle on peeking in from the outside. I could only spot one bat that made regular appearances in the entrance chamber. It displayed some rather strange behaviour for a bat; it was very inquisitive. While bats are curious animals by nature, when close to the roost or inside it, in the case of cave bats, they tend to tolerate only minimal levels of disturbance. A bit like how strangers approaching you in the park is acceptable for most people, the same can't be said about one's garden. But this particular bat was different; it flew back and forth between the entrance chamber, where I could see it and where it obviously could sense my presence, and the deeper parts of the underground system. One wouldn't expect a disturbed bat to do that.

Jesper helped me get my bat recorder from the car, but when I turned it on, I wasn't getting much helpful information from it. This led me to believe that it was a phyllostomid because of their very faint, and in this case unnoticeable echolocation. Things had got interesting: an inquisitive phyllostomid, of medium size, with no apparent markings on its body. The options started flashing through my head, but I kept going back to the same potential suspect – the Vampire Bat *Desmodus rotundus*. I didn't count it for my list because it was just a hunch; I had no actual evidence to confirm it.

The second cave, we never found. As we got to the location of the GPS coordinates I'd noted down, we stumbled onto one of the biggest flaws of relying solely on coordinates to find caves. We were most likely very close, but there was a drop of 200 m or so in front of us. One could assume the cave was at the bottom of that. I'll probably never know. It was a turning point in preparing for my future cave visits. We missed both caves because GPS coordinates often aren't enough for finding them. Once again, the locals saved the day and showed us another cave, which was full of bats, including the *Pteronotus* and *Natalus* species I wanted to find. Thanks to Jesper and his fluent Spanish, I was able to ask our guide about their relationship with bats, particularly with the vampire variety. As a farmer, I reckoned he likely had an opinion about them. He told us that they'd been working with researchers to develop an approach that includes testing the bats for rabies when they detected that their cattle had been targeted. This method is far superior to the far more common one in the Neotropics – that is, killing everything in the nearby caves. The main issue with that is that it's indiscriminate; all bats found are killed (often by burning the roosts), regardless of their species.

Yucatán, Mexico, July 2019

No trip to Mexico is truly complete without a stay in Yucatán, the most biodiverse state in the country. And no trip to Yucatán is truly complete without Juan Cruzado Cortes's expertise and knowledge of the local wildlife. He very kindly offered to show me around and let me sleep under his roof.

The first of our many visits to Maya temples was in Uxmal, for the Broad-eared Free-tailed Bat *Nyctinomops laticaudatus*. It is unusual to see them this close to the ground as they require some height to fly off, but I suspect that the temple's structure allows for that height while also offering roosting features 2 m off the ground. Two things betrayed their presence: their incessant squeaks and a dead pup on the ground. These things happen all the time in any roost; I didn't think this was due to the disturbance caused by tourists in the temple. I was astonished to see how close the bats and people were, probably without the latter knowing anything about it. It made me think about how long bats may have been using these temples. Two hundred years? Five hundred? If it's the latter, does that mean they lived alongside Mayas? Or did they only move in once the temples had been abandoned? The temple was also home to a small group of Jamaican Fruit-eating Bats *Artibeus jamaicensis*, a very common species throughout Central America; and, more surprisingly, a pair of motmots! Those giant bee-eaters are gorgeous birds displaying bright blues and greens, like the rest of the bee-eaters, but they're a lot bigger. And the inside of a temple is probably the last place I'd have expected to see them. And yet, here they were, mostly undisturbed by the presence of people, except when it became too loud.

The next evening, we met up with a group of ornithologists on their way to Colonia Yucatán, not too far from Cancún, for bird surveys. As we were sharing nets, we thought it would be easiest to go along with them and trap bats there in the evening and then give them the nets afterwards so they could trap in the morning. Our first trapping location was outside a cenote, a cave formed by the collapse of limestone bedrock, exposing groundwater. They're well-known diving features, especially in Yucatán. Unfortunately, I didn't have the time or the budget to give diving a go. It was all about the bats here.

The presence of bats was immediately evident because of how loud the chatter was. The first bat we caught was a *Myotis*, of all things. I was not expecting to be catching such tiny bats given the abundance of Phyllostomidae in the cave. And yet, our first visitor was a Hairy-legged Myotis *Myotis pilosatibialis* – formerly known as the Montane Myotis *Myotis oxyotus*, but they've since been split. *M. oxyotus* is mostly restricted to the Andes, and *pilosatibialis* to Central America. The next find was Seba's Short-tailed Bat *Carollia perspicillata*. Anyone who's trapped in the Neotropics has caught one, likely dozens, but I was very excited. To me, this was (almost) as exhilarating as any other new addition to the list. That's one thing I love about Big Years: all species account for one point. On the flip side, it would be easy to be underwhelmed by new species because there are more to chase, but I didn't feel that way.

Overhead, more bats were flying and I had my recorder on, as usual. I identified them as the Black Mastiff Bat *Molossus rufus*, a common molossid species challenging to catch because of how high they fly; and the Mesoamerican Moustached Bat *Pteronotus mesoamericanus* (formerly *Pteronotus parnelli*). The latter is a common species in caves, so I'd been confident I would see it at some point. The nets were busy, mainly with *C. perspicillata*. I was learning to recognise it at a glance, so much so that when a Common Big-eared Bat, *Micronycteris microtis* flew into the net, I already knew it was different. I had no clue what it was because I hadn't had enough time to study this hugely diverse family, so I had no idea what I was supposed to look for. All I knew was that this one had particularly large ears. It was a start. Luckily, I had Juan! He recognised straight away this was a *Micronycteris* because of the long ears. Still, the differences between the species are subtle, often involving looking at the fur to see whether it's bicolour or tricolour, or measuring a thumb or a foot. That evening was a proper introduction to the world of Phyllostomidae. These obscure identification features make you wonder how many different features early bat researchers looked at before finding the few criteria that worked.

One bat that didn't stand out to me, but that did stand out to Juan, was a Sowell's Short-tailed Bat *Carollia sowelli*. This *Carollia* is smaller than *perspicillata*, but it does look very similar – its fur is a shade warmer, but that takes practice to notice. Much of Phyllostomidae identification comes down to training, as I would learn over the following months. While the published criteria have been proven as reliable, they're not the only criteria, and experienced field bat biologists are often able to distinguish bats only by glancing at them, before they even really look at the details. Back at our hotel, before going to bed, which I was keen to do, I spent some time recording some bats. It was handy to have Juan nearby, so that I was confident that I was indeed recording the Wild Bonneted Bat *Eumops ferox* and the Southern Yellow Bat *Lasiurus ega*, two new species for the list. Overall, saying this was a productive night is a bit of an understatement. I added no fewer than eight new species that evening, a record for the Big Bat Year.

Our second night of trapping in Colonia Yucatán was a lot quieter. Juan had chosen a location at the edge of the forest, in a disturbed habitat. While this might sound counterintuitive, disturbed habitats usually have much higher densities of bats but lower diversity. This lower diversity may only become apparent after a few nights of trapping, as even in pristine forest, catching rare bats is unlikely, as their name suggests.

The night started with lots of critters crawling and flying around us. While putting out the nets, I found a giant millipede. While I am not too fond of centipedes, I do love millipedes. They're like the much more laid-back, friendlier cousin of centipedes (how many biologists have I offended with that?) – the latter wouldn't mind taking a bite off your head if they could. I let the millipede crawl all over my hand. It was mostly black but had bright orange rings and legs, a stunning creature! And when I say it was big, I mean compared to the one-centimetre millipedes I was used to seeing at home. This one was easily over fifteen centimetres, which probably isn't that large by tropical standards.

Our first bat of the night was a Cozumelan Golden Bat *Mimon cozumelae*, now called *Gardnerycteris cozumelae* – a rather striking member of the Phyllostomidae. As these few trapping nights were my first taste of Neotropical bats, I was only just starting to discover the incredible diversity of this family. *Mimon* has a very long nose-leaf, almost like a sword but not quite long enough to grant them the title of 'sword-nosed bat', seemingly restricted to *Lonchorhina* spp. It was also surprisingly colourful – all the Phyllostomidae I'd been catching so far were of various shades of brown. This one was almost bicolour with a pale belly and a darker, greyer back. The next one was even more striking, with prominent stripes on its face. Lots of Phyllostomidae have stripes on their faces, sometimes on their backs too, but when they're as noticeable as they are in the Pygmy Fruit-eating Bat *Artibeus phaeotis*, they look extraordinary. I won't get into the issue of *Dermanura* versus *Artibeus* for the genus, as I've been unable to find enough people agreeing on either one of them, including in publications, to know which side I should be on. It shouldn't be a matter of opinion, but it seems to have become one, as is sometimes the case in taxonomy.

In between the various bats we were catching, we also saw and extracted many bugs from the net. Most moths don't get entangled, but some of them do and so do beetles – their exoskeleton makes them a pain to extract without killing them. As things quietened, I took a break and sat down, away from the nets, torch off. I heard some noise; leaves were rustling nearby, and whatever it was, it was coming closer. I eventually decided to have a look at what it was, for safety reasons, if nothing else. It turned out to be what I thought was a coral snake, a highly venomous group of species but not at all aggressive. It turned out to be a false coral snake, but its mimic was definitely very effective on me. Despite this, the snake was far more terrified than I was and disappeared in a split second.

The last bat of the evening was a Common Vampire Bat *Desmodus rotundus*. It's a fantastic species to catch. Interestingly, this one was not nearly as bitey as other Phyllostomidae, bearing in mind this is a self-defence mechanism. Upon release, I was hoping to film it flying off, but it did what they often do, which is jump and run away. Such a bizarre behaviour for a bat – but vampire bats are well adapted to running, as their hind leg muscles show. There was nothing noteworthy on the recorder except the Argentinian Serotine *Eptesicus furinalis*, which was new but not unexpected.

The cutest little vampire, Hairy-legged Vampire Bat *Diphylla ecaudata*.

My bat list was growing at an incredible pace, and Juan kept suggesting more and more places we could go to expand the list even further. Not since Thailand had I seen it grow this fast. When he suggested a cave with the Hairy-legged Vampire Bat *Diphylla ecaudata*, I couldn't say no. This species was the inspiration for the bat's face on the Big Bat Year logo; never seeing it would have been a bit frustrating. It's arguably one of my favourite bats in the Neotropics too. For me, it was well worth the day trip. The cave wasn't too far from Mérida but going there still meant we couldn't do much else that same day. We got to the cave after a half-hour walk from the car. The cave entrance was small and the whole cave wasn't exactly spacious either. Well, there was volume, but the shape of the rock formations made it difficult for a human being to crawl inside, which is likely why it contained so many bats. At the entrance, we were greeted by the usual *Carollia* while the rarer species were further back. Juan almost bumped his head into a couple of bats; they were *Mimon cozumelae* again. I recognised the slow fluttering of *Desmodus*. Funny little creatures they were, hopping around the cave, only using their wings every so often. *D. ecaudata* was a far less active, yet far more inquisitive bat. It landed on a rock ledge only a couple of metres away from me, let me take a few photos and then left again. I thought I may have disturbed it with my flash but it quickly came back. It probably wasn't satisfied with its grasp of the situation and needed a closer look. Vampire bats have incredible personalities!

While we did a lot of trapping, we didn't trap every night – that would have been too exhausting, and it was also logistically a bit difficult as the nets were shared with the bird people. We drove around town to a Starbucks one night when we didn't have the nets. None of us were interested in their overpriced coffee. No – instead, Juan knew that the house right across the street had a roost of Alvarez's Mastiff Bats *Molossus alvarezi*, along with Yucatán Yellow Bats *Rhogeessa aeneus*. The latter species is endemic to the Yucatán Peninsula, so it was an excellent addition to the list. Most current *Rhogeessa* species have been described quite recently, following a series of splits. Their limited distribution range makes it easy to see a handful of species without having to travel very far.

The main goal of the trip was to visit the famous reserve of Calakmul, officially Reserva Biosphere de Calakmul, home to over a hundred species of bats. It was the furthest site from Mérida we visited, but it was the most exciting drive. Just as we entered the park, a cat crossed the road. You might think that's not that thrilling; it's likely a regular occurrence where you live. But Calakmul is a proper nature reserve – it doesn't have house cats. Those would get eaten by street dogs before they reached the reserve and should they reach it, the Jaguars would take care of them, as they do with dogs abandoned by people. This cat was special: it was plain, dull in fact, but it looked wild. It was one of the very few Neotropical cats to be active during the day, the Jaguarundi. That was a proper sighting of a proper cat.

We watched an emergence at a pit called 'El volcán de los murcielagos' – in English, 'the volcano of bats'. There were hundreds of thousands, if not millions of moustached bats. The total estimate is roughly three million, but those numbers always come with substantial margins of error. Unlike Bracken Cave, this cave had multiple species, mainly the Tawny Naked-backed Bat *Pteronotus fulvus* and *Pteronotus mesoamericanus*. Luckily, the calls of these species do not vary as much as calls from other families with lower duty cycle calls, so it is perfectly fine to use a recorder when there are thousands of bats flying out of a cave. Yes, the sonograms got a tad busy, but the characteristic frequencies were enough to accurately identify the species present. Had there been a Vespertilionidae of some sort in the lot, and I suspect there was, I wouldn't have had a clue what it was.

The road leading to the cave had a sign telling drivers to be mindful of bats. I'm not sure how helpful this sign is, but it was the first time I saw such a warning for bats. I'd already seen quite a range of signs for kiwi, penguins, turtles and snakes, but not bats. We were trapping at the edge of a pond in the middle of the forest. Unlike our other trapping nights, it was truly dark. There was no moon to be seen, no light pollution, just a pitch-black expanse where the pond was. We caught very few bats, but the activity on the recorder kept me busy. I was able to add the Black-winged Little Yellow Bat *Rhogeessa tumida*, a different species of *Rhogeessa* than the one I'd found in Mérida, as well as the Ghost-faced Bat *Mormoops megalophylla*. While seeing it would have been a lot nicer, recording it was still better than not being able to add it to the list at all. It's a species that many people think is really cute, the kind of thing that's so ugly it becomes attractive, but I really don't get it. It just looks weird. I think you could see its face upside down and not notice it's upside down, because of how strange it looks. I also recorded Underwood's Bonneted Bat *Eumops underwoodi*, yet another molossid for my list. That is a family I've reliably been able to record in most places around the world, probably because they're loud bats.

Mayan temples are home to many bats.

While Juan and I would have loved to see the Spectral Bat *Vampyrum spectrum*, he could not coordinate a roost visit with Rodrigo Medellín, a world-famous Mexican bat researcher. I wasn't going to do any valuable research on it. I was there as a tourist, which meant some things were off limits, and I was OK with that. I always put the animals first, especially if I knew they were sensitive to disturbance. That's the reason why I didn't visit a roost of the Australian Ghost Bat *Macroderma gigas*, for example. But Juan did want us to visit yet another temple, away from all the others. It would normally have been unreasonable, especially for just one species, but this one was special. Our target was the Woolly False-vampire Bat *Chrotopterus auritus*, *Vampyrum*'s smaller 'sibling'. Like *Vampyrum*, it's a mainly carnivorous bat with a diet based on large insects, small mammals, birds and other bats. It's a lot smaller than *Vampyrum*, though, and slightly more common. The temple we visited had an area cordoned off to give the bats some peace and quiet from the tourists visiting the ruins. But we weren't there to visit the ruins per se; we were interested in seeing the bats.

Juan went in first and saw that the bat was present. In an attempt to keep disturbance to a minimum, I made sure everything was ready for a couple of photos; I didn't want to bother the bat for more than thirty seconds or so. I took two photos and looked at the results on the back of the camera. They looked sharp, so we exited the temple. Outside, I looked more closely and saw that it had a pup. I counted three ears on that pup, which didn't quite add up. After another look, I understood there were two pups, a rare phenomenon in Phyllostomidae, especially carnivorous species. This proves once more that bat photography, when done responsibly, isn't all bad. We would not have seen the second well-hidden pup, were it not for the photo. This constituted the first record of twins for that species! However, after discussing this with Rodrigo Medellin, it appears that the two pups might perhaps not be related and that one would have been adopted. The plot thickens!

The last night of trapping that Juan had planned for us was in Mina de Oro, inside an abandoned building along the coast, on the hunt for Mastiff Bats. There weren't just bats there – we found the burrow of a tarantula and a python near the ceiling that seemed very interested in the bats resting on the walls. A large scorpion was snacking on an equally sizeable cockroach. There was a lot going on in that building. Not so abandoned after all. We were quickly overwhelmed with the number of bats we were catching, so we decided to close up the nets to avoid capturing the same two *Molossus* species over and over again; it all felt a bit fruitless at that point. The heavy winds also made the release of some

Twins or not? That is the question. Woolly False-vampire Bat *Chrotopterus auritus*.

of the bats rather challenging. Molossids can't really take off from one's hand as most species can; they need to be tossed in the air to be given enough height. This shows how specialised they truly are. While the vast majority of the bats we caught took off once they'd been thrown in the air, a few of them did come back down a bit too quickly, so we had to find them crawling through the vegetation and give them another go. This is the reason why their roosts are so hard to find; they have to be high off the ground, making a trapping site such as this building highly unusual – here we'd caught dozens with a regular net that was 3 m high. They appeared to be using the building as a night roost to rest between each course of their fancy evening dinner. On the way back to Juan's house, we stopped by a lake to record the Greater Bulldog Bat *Noctilio leporinus* and the Big Crested Mastiff Bat *Promops centralis*, which we got very easily – concluding this mind-boggling trip where I'd added 46 species to my list, another Big Bat Year record! My first time in the Neotropics had been a resounding success.

Vampires and people

Bat workers, myself included, sometimes describe bats in the hand as aggressive. I've done it in this book even. That's not entirely appropriate, as we're using this adjective to characterise a self-defence behaviour. Bats can't really be described as aggressive since they don't bite unless handled. I think it's pretty fair for any creature to bite when handled – it's not particularly fun to be restrained against your will. Handling induces stress in bats, which they express by defending themselves. As a result, handling bats should only be done when it makes sense from a research and/or conservation perspective. While handling bats can be fun, without a research question to justify it, it's irresponsible. Equally irresponsible is the poor handling of bats, for example by stretching the wings too far towards the back of the bat or by suspending it by its fingers. Those practices are sadly too common among bat researchers. Proper bat-handling technique, including wearing gloves, significantly reduces the risk of bites and in many cases, it reduces the stress experienced by the bats too.

Some of the most common questions that arise when discussing bat bites include: 'Aren't you worried about rabies?' and 'What about vampire bats? Do they bite people?'

Beyond the tales, vampire bats do, in fact, exist. The Common Vampire Bat *Desmodus rotundus* is a fascinating example of a species doing much better around humans than in primary forest. It is believed that their density was very low in pre-Columbian times, similar to the levels still seen in undisturbed habitats, away from human settlements.[1] This is because *Desmodus* feeds on the blood of large mammals. It's been recorded feeding on tapirs, brockets, sea lions and even penguins and pumas.[2,3] In the rainforest, its options are somewhat limited as the densities of large mammals were very low – until colonisers brought in cows, that is. The explosion of cow farming in South America has led to an explosion of *Desmodus* populations in many parts of Latin America. This has led to more common interactions between bats and cattle, and humans to an extent. While human–wildlife conflicts often find their roots in humans colonising nature; in this case, it seems that humans brought the bats upon themselves, increasing their population along the way.

Bats feeding on the blood of cows isn't necessarily an issue. Still, when the densities of vampire bats become so high that there are dozens of individuals feeding on a few cows every single night, the stress on the cattle is significant and can lead to disease and death. There is also the issue of proximity to human settlements, leading to the occasional human blood supper when people sleep unprotected by mosquito nets. While bats, including vampire bats, are far from being the primary vector of rabies, the increased densities of vampires and the subsequent increase in bites have led to a rise in rabies cases across parts of Latin America.

The impact of vampire bats on cattle and people have caused many outbursts of anger against them, resulting in cave roosts being decimated. These bat colonies are destroyed indiscriminately as most people cannot distinguish vampire bats from other, more common types. Vampire bats often roost away from the other species – or rather the opposite, the different species roost away from them – which makes them reasonably unlikely to be killed in these haphazard attacks.

Notes

1 Delpietro, H.A., Marchevsky, N. and Simonetti, E. (1992) Relative population densities and predation of the common vampire bat (*Desmodus rotundus*) in natural and cattle-raising areas in north-east Argentina. *Preventive Veterinary Medicine* 14 (1–2): 13–20. https://doi.org/10.1016/0167-5877(92)90080-Y

2 Luna-Jorquera, G. and Culik, B.M. (1995) Penguins bled by vampires. *Journal für Ornithologie* 136 (4): 471–2. https://doi.org/10.1007/BF01651597

3 Kays, R. (2016) *Candid Creatures: How Camera Traps Reveal the Mysteries of Nature.* Baltimore: Johns Hopkins University Press.

Antsiranana, Madagascar, July 2019

As I was booking my flights to the north of the island, I could hear the faint yet distinct cry of my wallet that I always hear when I make a terrible financial decision. The prices of domestic flights are utterly ridiculous, the main reason being, as always, tourism. Tourists are the only ones willing to pay significant extra cash for the sole benefit of getting to a remote location faster. In terms of prices, the sky is the limit, quite literally in this context. Unfortunately, flying was my only reasonable option in order to get far enough from Tana that I was guaranteed to see some different species. Given the prices, I had to have a high level of confidence in my chances of recording new species, if I didn't want this to be a waste of money. There was one rather significant issue, though. While I had coordinates for a couple of bat caves, I had no way of knowing how to get to them, and beyond that, I had nothing. As my adventures in Mexico showed, this was risky – my chances of actually finding the caves without any local help were slim.

The check-in office of my accommodation was located in town but the lodge itself was in the middle of nowhere. The lodge has a shuttle service but by the time I got to the office, the last car had already gone. That meant I had to wait a while for the lodge to send one out to pick me up. It came soon enough, and we were on our way. At some point, we drove through a large sandflat of some sort; I saw lots of bats flying around, so I asked for a quick stop so I could use the bat recorder and at least have a chance to put a name to them. They were Malagasy White-bellied Free-tailed Bats *Mops leucostigma* and Malagasy Free-tailed Bats *Tadarida fulminans*. The former could actually have been identified in flight because of its obvious white belly, but I'd decided against visual identifications.

As I arrived late at the lodge, dinner was already well underway for the other guests, so I had no other company than my own, which I'd got used to by that point. After dinner, one of the staff members took me for a short walk around 'Jungle Park' to help me record some bats but also to show me some cool critters. He found a few chameleons, which I would have missed entirely, and a few nocturnal geckos too. Bats were flying around as well, including the

Madagascan Straw-coloured Fruit Bat *Eidolon dupreanum*, an endemic closely related to the African Straw-coloured Fruit Bat *Eidolon helvum*, a migratory bat species currently holding the world record for the largest mammal migration. *E. helvum* is a widespread species in continental Africa that sometimes forms colonies of millions. Those large concentrations make the species vulnerable to threats such as habitat loss and persecution, particularly in cities. Yet it was one I was guaranteed to see in Kenya. Scoring its cousin was much less of a certainty, so this sighting was definitely more than welcome. The evening ended with an utterly adorable Northern Rufous Mouse-lemur, a highly localised species.

The following day, despite being ill for the first time on this journey, I arranged for a guide to take me to a cave that seemed likely to contain bats. I had no way of knowing whether this was the cave I'd found coordinates for, nor whether it would actually have any bats – but I thought I might as well give it a go. While we saw a few lovely birds and some rather beautiful snakes, it was the baobabs that caught my eye. I'd seen a few before, but I'd never seen a forest where they were the dominant tree species—combined with the rocky terrain, that made for genuinely breathtaking scenery. The cave contained Western Sheath-tailed Bats *Paraemballonura tiavato*, and nothing else, unfortunately. I was hoping to find the Rufous Trident Bat *Triaenops menamena*, as that was the species I'd found the

The famous baobabs.

cave coordinates for – but this cave didn't seem to have any. It just so happened, however, that this species had been waiting for me at the lodge; I found them foraging above the tiny pond in the evening. Turning on the EMT while I waited for my food to arrive got me *Triaenops* and the Malagasy Myotis *Myotis goudotii*, as well as a couple of long-fingered bat species. I'd picked the lodge based on its proximity to caves, though I wasn't sure I could access them. I hadn't considered the fact that this small artificial body of water would also be attracting bats from all around.

Andasibe, Madagascar, July 2019

The same taxi driver who'd driven me to my hotel on my first day on the island drove me to a park on the east coast. On that first day he'd told me he could take me to Andasibe to see lemurs if I wanted. He was charging a reasonable price and was friendly, so I had no reason not to take him up on the offer. The east coast has a different bat fauna to the west coast, which is very similar to what I saw in the north. A whole new selection of *Miniopterus* and *Scotophilus* species was therefore available to me. We had a fairly lengthy journey ahead of us, although perhaps not by Malagasy standards. We were headed for Andasibe National Park; it was time to see some proper lemurs. I know, I know, mouse-lemurs like the one I'd seen in Vallée des Perroquets are cute and often rare, but I wanted to see some of the more charismatic species. You know what they say: 'When in Rome...'

While they're not as well known and perhaps not as charismatic as, let's say, Ring-tailed Lemurs, I was keener to see sifakas, indris and whatnot than the 'generic' lemurs that everyone thinks about when they hear the word. I was headed to Andasibe and not anywhere else because this place is very accessible from Tana. The drive is under three hours, on decent roads, no expensive planes required. Very few interesting places on the eighth continent tick those boxes. It was a bit of a no-brainer if you ask me.

Along the way, besides the few birds I saw, I also realised how much of the natural habitat has been destroyed. It's no wonder most of the wildlife is now restricted to a few pockets of protected forests scattered around the island. Quite the depressing sight; even more so than the palm plantations in Malaysia, I'd argue. It of course isn't a competition, but Madagascar stands to lose a lot more unique wildlife than Malaysia right now. Both are horrible situations that need to be addressed sooner rather than later. As we got to the hotel I'd booked, I heard there were regular night walks in the vicinity of the park for various wildlife species, from reptiles to mammals. For me, this would also be an opportunity to walk around with my bat recorder. Unfortunately, I didn't have time to arrange anything on the night of my arrival, so I planned to get myself on one of these walks the very next day. That evening, I walked around the hotel grounds and recorded the Malagasy Serotine *Neoromicia matroka* and Major's Long-fingered Bat *Miniopterus majori* – not a bad start, but I was hoping for much more once I got closer to the protected area.

The following day, the first thing we did was head to the village to figure out how to get on one of these night walks. However, as I asked around and was directed to Marie Lodge Andasibe, accommodation on the main road leading to the park, I also heard there was someone in the area who knew all about bats. I had to find that person. It took some back and forth, but in the end, I met Daniel, the local batman, who told me his son would meet me that evening, and we'd go from there. I was excited. I was effectively getting a private night walk with some mist-netting as a bonus too.

After that, I headed to the national park. At the beginning of the guided walk, we focused on birds, and we saw several endemics, especially around a pond that seemed to have quite a lot of activity. The highlight for me was the Blue Vanga. Vangas are a group of birds endemic to Madagascar, closely related to the helmetshrikes from continental Africa, except vangas come in stunning colours. I did not get to see the famous Helmet Vanga, but I'll save that for another time. As in many other places, I missed many common and rare birds, but that was simply because my focus could not be on every animal simultaneously. The pond would likely be good for bats too, but it's not accessible to the public at night. Even the night walks don't enter the protected area itself.

A little further on, we stopped as we met another group that had spotted a couple of Diademed Sifakas moving in the trees. We got our eyes on them shortly after – they weren't exactly discreet, so they were easy to spot. They're incredibly colourful lemurs with a mix of light orange, white and various shades of grey. I think they're one of the prettiest species out there. We followed them for probably ten minutes, as they did not seem bothered by our presence. As time went on, and the news spread of the sifakas giving a good show, more people came, and it quickly became a bit crowded for me. I decided to move to a quieter area while retaining a view of the lemurs – not as close as I would have liked for the pictures, but it was becoming uncomfortable. The lemurs must have read my thoughts somehow because they came straight towards me. While chasing each other, one even jumped right above my head, so close that I could feel the air move. I got fantastic pictures, both with my DSLR camera but also with my smartphone when they were too close for my lens to focus. Crazy! I'm not a fan of approaching mammals for pictures, but when they come straight at you, there's absolutely nothing wrong with making the most of the situation. That is precisely what I did.

After spending another twenty minutes with the lemurs, who still seemed unbothered by the many tourists, our guide asked if we wanted to see an indri. After careful consideration, we all agreed. Why wouldn't we want to see indris? We got to the indri tree after a walk of only about five minutes. Yes, this part of the park was small, and everything was close together; it was a bit surreal to see so much diversity in so little space. There were already a handful of people watching the indris and trying to get clear photos of them high up in a tree, somewhat of a challenge. Like the other guides, ours tried to direct photographers to what they considered the best shooting spots. Not only did those get very crowded, very quickly, but I actually found a spot that I thought was better. Sure, I had to get a bit creative with the exposure, because the light wasn't great. But eventually, I got photographs I was happy with, without pushing anyone to get them.

I guess these two sightings perfectly illustrate the issue with easily accessible parks and charismatic species; they attract crowds and lots of them. It's a good thing these animals didn't seem to be bothered at all, but in other parts of the world, those crowds can become problematic. But they also help fund the conservation of those reserves. Well, that's the goal anyway. It's hard to know how effectively each reserve uses the money they collect from tourists. This experience reminded me a lot of the tarsier excursion in Tangkoko. Still, neither the tarsiers nor the lemurs seemed to suffer from the attention, and the funding they help generate most likely helps protect many other, far less charismatic and cute species in the area, including bats.

I had pretty much seen everything I could, and I was getting tired of the crowds, so I headed back. Once back at the hotel, I had the nasty surprise of discovering my website was down because of an update that had gone sideways. I quickly started to panic as I realised I did not have the skills required to solve the issue. After a cry for help on Facebook, I found someone who saved the day

One of the prettiest Lemurs out there, the Diademed Sifaka.

and agreed to guide me through the necessary steps to solve the problem. Social media can be a drag sometimes, perhaps more often than not, but in circumstances such as these, it can be an incredible resource to get help quickly. Most of my afternoon had flown by, by the time I'd sorted the issue, and it was soon time for dinner and bats!

Daniel's son, whom I'd arranged a night of trapping with, met me at the hotel's reception, and we left for the park again. Our first stop was the post office, but only because its attic was rumoured to be a bat roost. We set up a somewhat dubious-looking mist-net using various poles attached to each other in order to reach the height required to block the exit that the bats reportedly used. Unfortunately, it started raining shortly after we'd set up the net, and we were forced to pack up. We decided to go on a walk around the area. Instead of a guided night walk with a dozen other people, I was having my own private one. I kept the recorder running for as long as I could, between the rain showers, hoping to avoid another drama like the one I'd experienced in Australia. While my guide couldn't help me much with the acoustic identification of local bats, he was very knowledgeable about a lot of nocturnal taxa, especially chameleons. I was in luck because I've always found them fascinating. We also found a mouse-lemur. This one was a Goodman's Mouse-lemur but it did closely resemble the one I'd seen up north. Before it was time for me to head back to Tana for my flight to Nairobi, he showed me a house that had bats inside. They turned out to be the Robust House Bat *Scotophilus robustus*, my 15th species in Madagascar. I'd been incredibly lucky in finding so many. Those 15 extra brought the total to 253.

Nairobi, Kenya, August 2019

When we woke up, we wondered where the sun had gone. And the warm air. And the entire city, for that matter. All we could see was cold, thick, wet fog. Manu had once again joined me for this leg of the journey and he was disappointed to find the kind of weather you'd expect in, let's say, Scotland rather than Kenya – and yet, there we were, quickly realising that opening the window first thing in the morning when it was a mere 8 degrees Celsius (that's 46°F for our imperial friends) outside wasn't the brightest idea we'd ever had. We also had no incentive to stay there any longer, so we packed up and left the flat to find the bus station, in order to head to Samburu in central Kenya. After asking around at the bus terminal, we quickly located our bus and on we went.

We knew we'd have to stop in Nanyuki to hop onto another bus; we thought we'd use that time to grab a bite, which we figured we'd need after a five-hour drive. In town, we entered one of the very few restaurants near the bus stop. We glanced at the menu – nothing unexpected there, it was typical Kenyan restaurant food. So we ordered some fried chicken and chapatis. It turned out they didn't have that. So we ordered something else with chicken. It turned out they didn't have that either. Realising we could be here all night if we carried on like this, we asked what exactly they did have from the menu. They had one dish only, a mixture of potatoes and meat. It was a tasty dish but it wasn't even on the menu and it would have been even better if we hadn't wasted time trying to order things that were on the menu but not available.

Having filled our stomachs and bought a couple of water bottles, we went looking for our second bus, which we found even faster than the first one. The fact that only two buses were waiting may have helped. Another two hours on the road and we reached Archers Post, the town outside Samburu National Reserve. I'd decided to spend some time there for a short safari before eight days of intense batting with the famous Kenyan bat researcher Paul Webala. It also happened to be our only shot at a much-wanted species, the Yellow-winged Bat

Lavia frons. *Lavia* is one of the very few bat species reliably spotted and identified by birders, where it is present, as it's often found roosting in bushes or flying out of them after being disturbed.

The 'action' was only supposed to start the next day, on my birthday, when we entered Samburu, but we agreed that a walk around the hotel with a bat recorder couldn't hurt. We headed to the outskirts of town, after what we knew would become a painful ritual of battling with the most bizarre door-locking mechanism either of us had ever seen. It's fair to say we didn't get very far, as the wasteland we wanted to have a look at really didn't seem too appealing. I was recording some bat activity, but I wasn't sure what was present. I was getting some molossid-type calls as well as some steep frequency-modulated calls. We decided to have a look at the other side of the hotel – maybe that would be more appealing or have more bat activity.

It wasn't. Until it was. I saw a yellow flash in the corner of my eye; my first thought was that something had flown in front of a yellow light and tricked my brain. My brain gets easily duped when it really wants to see something. *Lavia* wasn't yet at the forefront of my mind, but it wasn't very far away. When I checked said light, it was a cold white LED, the sort that would never cause a yellow flash in a million years, unless there was something yellow in front of it. Now *Lavia* really *was* on my mind. I couldn't figure out what else it could be. Another glance at my smartphone running the bat recorder, and I knew what was going on: a *Lavia* was hunting in the area. We HAD to find it. A minute hadn't even passed before we did. Or, more accurately, it found us – it went to perch on the roof right behind us. After confirming that this was the bat I'd been recording all along by pointing the recorder directly at the animal, it quickly became apparent that this could make for a great photo opportunity. There were plenty of bright streetlights around that didn't seem to bother the bat, so I assumed a few photographs with a flash wouldn't hurt.

I ran to the hotel room, battled the lock again, emerged victor, grabbed my camera and flash, battled the bloody lock again, won once more and ran back to the bat. While my last two minutes had been busy, Manu's and *Lavia*'s had been much quieter as neither of them had moved. *Lavia* gave me plenty of time to get some decent shots, adjusting my settings along the way, not at all bothered by the flash. We eventually left the bat and returned to the hotel for a good night's sleep. We had another opportunity to see the species inside the park, two of them in fact in broad daylight, but the highlight of our short stay in Samburu was probably the Heart-nosed Bat *Cardioderma cor* that I had the pleasure of sharing a bathroom with. Most people probably wouldn't agree that the only thing to make a shower perfect is the addition of a bat, but most people aren't doing Big Bat Years either.

An early birthday present, Yellow-winged Bat *Lavia frons*.

Shower buddy, Heart-nosed Bat *Cardioderma cor*.

Back in Nairobi, Paul, the bat researcher from Maasai Mara University we'd spend the next eight days with, couldn't believe we'd walked from the bus terminal to the hotel. Evidently, this wasn't something people normally did and the looks we got along the way were further evidence of that. Paul had prepared a schedule packed with lots of cave visits; most of them he knew well, and others he wanted a refresher on, to see if he could take along the students from the Global South Bats course that was due to start a few months later. We spent four days south of Mombasa and four days north of it, combining as many cave visits and trapping sessions as one could reasonably fit in a week. Hint: it's a lot.

Our first stop was a small cave on the edge of a town. We were in a rush as the sun was about to set, and obviously, we wanted to catch some bats before they emerged. Paul needed to collect reference calls for many of them, which was done by releasing the bats outside the cave in broad daylight so we could follow the bats around with a microphone easily. Usually, they looped two or three times before returning to the caves, which produced perfect free-flying bat calls, all without a single case of predation by a passing falcon or some other hunter. Quite an effective method! Doing the same at night requires some kind of light tag, which can be difficult to source and heavy for the smaller bat species. Any form of containing the bats, be it using flight cages or zip-lines, is unlikely to produce calls similar to those found in the field, making those practices ill-suited for most species. Recording these bats in the daytime makes following them with recorders much easier and, because they weren't taken too far from their roost, they weren't really in any danger of predation. This way, no light tags are needed. Using this method, we were able to record the African Trident Bat *Triaenops afer*, African Sheath-tailed Bat *Coleura afra*, Giant Leaf-nosed Bat *Macronycteris gigas*, and many others.

While most of the locations we visited were filled with bats, sometimes with millions of them, there were less successful locations too. One had been given to us by a couple of locals in the villages we were staying at and turned out not to have any bats. According to Paul, this sort of tactic was part of a common scam, which is why he'd told them he'd only pay them if there were bats.

It wasn't all about the caves; there was trapping too. For poles, we decided to chop down some bamboo. We stopped on the edge of a reserve where Paul knew we'd be able to find some sizeable bamboo trees. The place was also known for Golden-rumped Elephant-shrews, which Manu and I looked for. Our lack of success was probably due to the fact that it was the middle of the afternoon, likely the worst time of the day to see most mammals, particularly elusive ones. Keen to try some of the chopping myself, I asked Paul for the machete and tried my best, failing miserably to cut even just one bamboo pole. It's a lot harder than it looks. My failed attempts made Paul laugh, and from there

The author, extracting a bat from a mist-net.

Caves filled with millions of bats.

onwards, I was named 'City boy', for obvious reasons. One thing I did manage to help with, however, was the carrying of the poles up to the van, where we tied them to the roof. Those poles turned out to be incredibly productive on the two nights we used them, as we caught the Egyptian Slit-faced Bat *Nycteris thebaica* and Trujillo's Yellow Bat *Scotophilus trujilloi*. Both were new species for me. I was getting very close to 300 species. We also almost caught a motorcycle in one of our nets, which would have been a disaster. I'd spotted a headlight in the distance headed our way. Before I had time to tell Paul of its approach, the driver had turned off his light. Paul and I stood in front of the net in order to prevent the driver from driving through it, but he really wasn't keen to stop! When he understood we had no interest in his business or shady behaviour, he obliged and carried on, leaving our net unharmed.

I like to believe that the success of our trapping was entirely due to the input I had in carrying those poles to the car. I can't see any other factor that could have come into play. It's worth noting that I handled none of those bats and left them entirely to Paul, as both species were extremely angry, and angry bats and I don't do well together. The last species for Kenya, number 284, came in the form of Andrew Rebori's Yellow Bat *Scotophilus andrewreborii*, which we were able to record by making a brief stop on our drive back to Nairobi airport.

St Lucia, South Africa, August 2019

I had come back to a place I'd seen before, which hadn't happened since early January when I was in New Zealand. It gave me a feeling of comfort; everything felt familiar, almost like home. I stayed at a hostel in town, hoping to join a couple of night safaris while over there. I spent my first night in a tent set on the second floor. It's a bit odd to be sleeping in a tent in a building, but it was also an easy way to bring privacy to the place so that they could offer more than just dorms. I spent the first evening walking around the hostel grounds, hoping to detect some bats flying by, and so I did; I recorded the Hottentot Serotine Bat *Eptesicus hottentotus*. It's always oddly satisfying to record new species right where I'm staying, probably because it doesn't require me to drive for an hour in the dark to get to a place. A group of French students had arrived for a school trip of some sort. I have to admit that a safari-style school trip would have been a lot of fun while I was in secondary school – we did not have any of those. Shortly after dinner, things became exciting as the hostel staff informed us a hippo was wandering the streets. There are signs all over town warning of hippos at night, and everyone's reaction to this one walking past the hostel was to get as close as they could. People are weird sometimes. It entered someone's garden, probably to their despair because I seriously doubt any plants would survive a savage assault by a hungry hippo. The humans went back to their own business.

The north-east of KwaZulu-Natal (KZN) has an abundance of game reserves of various sizes that probably all have good bats. Hluhluwe-Imfolozi, one of the largest and most popular, likely has good bat diversity, but I knew actually finding any there would be challenging. The camps are fenced off, night drives are noisy and crowded, so I had no easy way to find bats at night. I wouldn't have more opportunities during the day either because, as is usually the case in popular areas, there isn't much room to accommodate strange requests. Clearly, this wasn't the best place to go. Because I had been to South Africa and KZN specifically before, I knew exactly where would be better:

Bonamanzi. It's a small, private reserve that doesn't get much traffic because while it's right next to Hluhluwe, it doesn't have any of the big cats people are so keen to see. Not having large cats and predators in general means the camp doesn't have to be fenced off; there's even a pond behind the restaurant, and quite importantly for me, it's not overrun with tourists. I booked a room there, sadly not in the treehouses my dad and I had occupied on our first trip there because those were well beyond my budget. In fact, a stay in Bonamanzi was beyond the budget I'd allocated for my South African leg, but I knew I would have good opportunities for batting, so I felt it was worth it. A two-night stop there wouldn't hurt my finances all that much. Shortly after booking my room, a couple of game drives and a boat trip on the river, I got an email saying they had to cancel all boat trips because there was a large male hippo with a snare around the neck, left by poachers. The park's staff were trying to remove it – but removing a snare from the deadliest mammal in Africa is no small feat. As a result, in the meantime, access to the river was forbidden for obvious reasons. The boat trip was therefore replaced with an evening drive, not quite the same. I had fond memories of this boat trip with my dad and our guide, who had turned out to be from Verviers, in Belgium, of all places. We saw lots of birds at close range, and because it was an evening cruise, we saw some bats too. In 2014, my knowledge of bats, especially South African ones, was minimal. As a result, I didn't know what I was looking for, but the prospect of doing this again had intrigued me. Sadly, a hippo made angry by poachers decided against it. Luckily the hippo was rescued in the end, from what I heard.

I saw nothing of note bat wise on the game drives themselves – not that I expected anything; I just wanted to see some animals. I also wanted to have another go at birding there as I knew I'd missed some good ones last time around. My main target was the Narina Trogon, Africa's only trogon, which I hadn't seen the first time. It did not take me long to find one – we got one on our first drive around the park.

Not having much planned for this leg of the journey allowed me to rest a bit while still adding new bat species to my list: the ideal scenario, if you ask me. I knew that my best bet would be the pond behind the restaurant. It was far enough that it wasn't very well lit, which is why I expected lots of bat activity. Any water feature in warm and dry climates is always attractive to bats and many other species. I did remember however that there were crocodiles in that pond, and in the dark, the likelihood of me seeing them wasn't very high. I'd seen how quickly they can disappear, even in shallow waters, in a neighbouring pond a few years prior, and I wasn't willing to try my luck. I made sure to follow the Australian rules I'd learned over there, always staying at a distance equal or greater than 1.5 times the size of the biggest croc. In this case, the crocs weren't

Lots of crocs, lots of bats.

that big, so 5 m or so would have been safe, but I stayed about 10 m away, just in case. Ten metres from the bank is more than close enough to record bats flying around it as well as foraging above it. Sure, I wouldn't record the bats hunting right in the middle of the pond, but I also vastly reduced the chances of becoming a croc's dinner right after having had my own, rather lovely meal.

As I expected, the bat activity there was sky-high; I'd made a gamble, yet again, but this time it was a bloody good one. Based on the information I'd been given, I knew I could find a different *Scotophilus* than the one I'd seen in St Lucia, as the African Yellow Bat *Scotophilus dinganii* is more of a coastal species in KZN. I had good chances of finding the Eastern Greenish Yellow Bat *Scotophilus viridis*, its inland sibling, and so I did. Above the pond, I also recorded Bocage's Myotis *Myotis bocagii*, which I'd ticked off in St Lucia too. Still, these recordings comforted me in my identification. It was also more exciting to record it in its typical savannah habitat than in the garden of my hostel in St Lucia. Sure, finding bats where I slept was fun, but finding them in their typical habitat was better. As I headed back to my room, I got stuck in traffic. Sort of. A herd of Impalas has decided that the area between my room and the car park was a good spot to spend the night. I tried to explain to them that they weren't wrong; they'd just come a bit too early. My attempts to get to my room scared them all, even though I tried to move as slowly as possible.

The KwaZulu-Natal Bat Appreciation Group got very excited when they heard I was coming. I just want to point out the name. Instead of the usual 'working group', it's an appreciation group. I think this certainly makes it appealing to more people, which is great. They offered to organise a trip to Shongweni Dam, just outside Durban. The main targets were Bushveld Horseshoe Bats *Rhinolophus simulator*, and the Natal and Lesser Long-fingered Bats *Miniopterus natalensis* and *fraterculus*. While catching the *Rhinolophus* would be easy, catching the other two wouldn't be. They were flying close to the ceiling, nearly eight metres above ground. Luckily, they can be distinguished based on their recordings, meaning I still had a chance at including all of them on my list. After we got through a tunnel in a rather worrying state of disrepair that somehow still stands despite having been built in 1925 with clearly very little maintenance, we came out the other side alive, and we entered the tunnel that contained the bats. We caught a bunch of *Rhinolophus* very quickly putting them in small mesh cages. These were initially intended for butterflies, but they gave the bats plenty of room to move around and hang from the ceiling, which the Rhinos did pretty quickly. The long-fingered bats, which we put in those cages shortly after, weren't so keen. I thought these mesh tents were a brilliant idea to keep the bats relatively comfortably while they waited to be processed.

Sharron, a bat carer based in Johannesburg, had invited me to see some different bat species than the ones I could find in KZN. Of course, I gladly accepted. I hadn't been there yet, whereas I'd already seen a good chunk of KZN, and it also wasn't a place I'd have wanted to visit on my own. It's challenging to overlook this city's bad reputation in terms of criminality. Sharron runs one of the very few bat rescue operations in the country and like other bat rescuers I'd met before, she'd transformed the entire house to focus on the bats' comfort. The garden had been turned into a series of flight cages to prepare the bats for release. One of those cages had an opening that allows for the soft release of bats. The creativity of bat rescuers never ceases to amaze me.

The Cradle of Humankind, as it's officially known, is a World Heritage Site region where some of the oldest human ancestor fossils and tools have been found, including Lucy, *Australopithecus africanus*, a famous, relatively complete (compared to others) three-million-year-old fossil; and Tumai, a seven-million-year-old hominid. A few years ago, 1,550 pieces of *Homo naledi* from around 15 individuals were also discovered there, becoming one of the most significant finds around modern human evolution this century. As I'd learned a lot about those fossils and their significance in our understanding of human evolution, I was keen to visit the museum there to see some of them in person.

As you might expect, the many caves aren't just home to many fossils – they're also home to many bats. Sharron had intimate knowledge of many of the caves and she was able to show me a couple where we found Temminck's Myotis *Myotis tricolor*, and the Natal and Greater Long-fingered Bats *Miniopterus natalensis* and *Miniopterus inflatus*. The latter is a large member of the genus, significantly larger than *natalensis*, but also comes in a bright orange livery! It's hard not to love a bright orange bat. Looking at the list of species I could still tick off in the area, two stood out: Percival's Short-eared Trident Bat *Cloeotis percivali* and the Rusty Pipistrelle *Pipistrellus rusticus*. They stood out for very different reasons – the first one because of its rarity and the difficulty in accessing their roosts, and the second one because of how common it is and therefore how bizarre it was that I hadn't seen it yet. We decided to go in search of the *Cloeotis* first because it would likely require the most time. Sharron knew of a mine that was a known roost. While the access to the mine is closed, thanks to illegal gold-mining, our hope was that we would have a chance at recording one emerging from it. Access wasn't the only challenge there, however; *Cloeotis* is famous for its high echolocation frequency, well beyond what most recorders can record. My only hope was that I would be able to record the first harmonic/fundamental of the call, which is usually present even though this species, like most Hipposideridae, tends to put the most energy into the second one, well above 200 kHz in this case.

When we got to the mine, the landowner warned us of the presence of armed miners inside the complex. This wasn't very encouraging news. While this could have been the best bat quest story, we were unable to find *Cloeotis*. The absence

Dracula in the Cradle of Humankind, Geoffroy's Horseshoe Bat *Rhinolophus clivosus*.

of the second species from my list, *P. rusticus*, stood out as it's a common one, and therefore one I should have come across it earlier. Before it was time for me to leave South Africa, and the African continent, we looked for a place to remedy that. Sharron's suggestion of a park not too far from the airport where we could stop before my evening flight turned out to be a good bet. I was well in time for my flight, having bagged the 60th species from Africa, bringing me to a total of 297 species overall.

Taxonomy

Taxonomy is considered by many to be a dying branch of science, something of bygone eras of the likes of Wallace and Mayr, frantically describing species and trying to figure out the relationships they have with one another. However, it is a forever evolving science, and this can be confusing or sometimes lead to heated arguments, as birders will know. But the field needs to reflect our developing understanding of the diversity of the natural world. Bats are in a taxonomic golden age, with over 200 new species added in the past decade – that's a lot more newly described species than with birds, for example, and more are being described every month.

Taxonomy is very important for conservation, as efforts need to be prioritised; not every species can receive the same level of attention. If choices need to be made, we need an accurate understanding of a species' risk of extinction. One 'species' once believed to be widespread, for example, was Common Bent-wing Bat *Miniopterus schreibersii*, with a massive distribution spanning from Europe to Australia, including Africa and much of Asia. However, it turned out to be a complex of species, including the 'true' *M. schreibersii*, which was actually in a relatively poor state of conservation. Another recent example from Europe also highlights the conservation implications: *Myotis crypticus* versus *Myotis escalerai*. Both species have recently been separated from the widespread *Myotis nattereri* that most European bat workers will be familiar with. While superficially similar to *nattereri*, these two species differ significantly in their roosting habits. While the *crypticus*, like *nattereri*, tends to use trees as summer roosts and only use caves in the winter, *escalerai* appears to be using caves as summer roosts. Their conservation requirements in the summer are very different, and without proper taxonomic description and extensive studies across the range, it's complicated to produce meaningful action plans. This is particularly challenging in poorly researched areas such as the Global South.

Why does it matter? A common question in conservation is 'What species use this particular site?' The answer can change over time because species may colonise the site or be extirpated from it, and because new species are always being described. The emphasis on collecting specimens is likely a contributing factor to the poor reputation of this field of science.

However, whole specimens euthanised for the sole purpose of ending up in a museum is no longer the only way, as wing punches can now provide sufficient genetic material in many cases. Without this, as is too often the case with bats, species lists are incomplete and include putative identifications based on old descriptions that don't work with currently defined taxa. This represents a challenge when species have to be assessed for the IUCN Red List for example, since in many regions around the world, most references might be a few decades old, meaning that the author's species list may be radically different from the current one. This makes it difficult to compare information at varying points in time, in order to pick out trends.

Mahé, Seychelles, November 2018

I had the opportunity to visit Seychelles for three weeks, only paying for my flights. That was an opportunity I couldn't miss. It came to be because my girlfriend at the time, Ellie, was conducting bird research there, and the team had a whole house rented so I could easily join them. It was the first time I'd be travelling specifically for bats, which was an opportunity to put my skills to the test, both in terms of finding and identifying bats in the field but also in terms of researching how and where to find them. Evidently, this wasn't quite the same level of challenge I'd encounter in Malaysia, for example, as the species diversity in Seychelles is limited. Only two species are known from the main archipelago group: the Seychelles Flying Fox *Pteropus seychellensis* and the Seychelles Sheath-tailed Bat *Coleura seychellensis*. Aldabra, a remote atoll, has a few more but we wouldn't have the time or the funds to go there, despite its appealing wildlife. As the bat diversity wasn't enough to keep two biologists occupied for three weeks, we also tried to find some of the rarer endemic birds, such as the Seychelles White-eye and the Seychelles Paradise Flycatcher, the Seychelles Parrot and the Seychelles Scops-owl. We were successful in the first three. We were not able to locate a scops-owl, unfortunately. However, the highlight of the trip for me, in terms of birds, was the Crab Plover – I'd wanted to see this very odd shorebird for quite a while.

Ellie had been seeing the flying foxes regularly, which is why even before I got to the island of Mahé, I knew it wouldn't be a challenge to find that one. In fact, as soon as we got to the house, I could hear them. I had to wait until the following day to see them, however. The second species was the real challenge, as finding information on *Coleura* was difficult. It seemed that very few people had been looking for it, and as a result, recent information was hard to come by. I'd found a few reports on observation.org, and given that it's the only 'microbat' in the area, confusion wasn't very likely. By far the easiest way to see the species is to throw money at the problem and book a stay at the Hilton hotel on Silhouette. However, as neither Ellie nor I had won the lottery and I had a world tour coming up that would chew through all my savings, this option wasn't available to us. The second-best option, and also the only alternative, was to visit Port Launay, in the north-west of the island. I'd heard of this location from an ageing paper on the distribution of bats in Seychelles. *C. seychellensis* is a critically endangered species, declining rapidly because of habitat loss, introduced predators and disturbance. These are the three main threats for any bat, anywhere in the world, but are usually exacerbated on islands. Because of all that, if I was going to have any chances at all of finding one, I needed pretty recent information on where to look for it. Paul Racey, a famous British bat researcher, had done some research there a while back, so I emailed him for some advice. He mentioned the boulders near Port Launay and gave me some specifics on the kind of crevices the bats looked for. He also emphasised the fact that this species is highly susceptible to disturbance. Because of this, we tried to limit my roost searches. Well, that and the fact that there was always a risk of ending up face to face with a giant centipede, which wasn't a very enticing thought.

I knew I had much a better chance of finding the bats in the evening on the recorder. As it is the only echolocating species there, identification wouldn't exactly be challenging. The bats emerged pretty late, or at least we first detected them quite late, despite focusing all our attention on that since that was really the goal of our trip to this part of the island. I wasn't too sure what kind of call I should expect as I wasn't able to find any reference calls, and it was the first time I'd encountered a species from the Emballonuridae family. While the calls of *C. seychellensis* are nothing special for the genus or the family, for a European bat researcher they were a novelty. We ended up spotting three distinct individuals foraging above the road, accounting for about 5–10% of the island's population, depending on the estimates. It's not every day that one can see such a significant proportion of a population, but that's also a sad realisation. As I was one of the first people to look for the species with a full-spectrum, I ended up making some of the first reference recordings. This has significant implications for conservation, as surveying the island using acoustics to find other remnant populations is a lot more manageable than setting up mist-nets, especially given that it's not an easy species to catch. The relative ease with which I was able to meaningfully contribute to our knowledge and potentially to future conservation projects (foreshadowing my return to the island in 2023) made me want to focus intensively on acoustics during my world tour. I would try to find as many species as possible for which I could make some of the first reference recordings.

While out on a diving trip shortly after, we stopped for lunch in the Cape Ternay area, not too far from Port Launay, on a part of the island that's only accessible by boat. I thought checking out the rock formations and crevices was worth the effort before heading back for the afternoon dives, as I doubt anyone had checked them for bats before. And the good thing about being on a diving trip was that I was already carrying a torch. No bat recorder, but that would have been of minimal use either underwater or inside crevices. This additional effort turned out to be unsuccessful but good fun, and a good reminder that not every attempt to locate rare bats is likely to succeed!

The quest for the world's rarest bat gave me lots to think about for my Big Bat Year. While I had decided not to come back to Seychelles during the world tour

Bat watching in the most amazing light, Seychelles Fruit Bat *Pteropus seychellensis*.

for financial reasons, this first experience travelling to remote places for bats was enlightening. The thrill of chasing a rare species in a relatively limited area, for example an island, is something I truly fell in love with. This likely explains why I spent the first month of my Big Bat Year on small islands, despite their limited diversity. Looking for rare species isn't just fun; it's also an incredible opportunity to raise awareness on some of the most severe threats bats are facing; and given that over 25% of the world's bat species live on islands, this represents quite a significant number of them. Island bats also often have fascinating evolutionary histories and are too often understudied. All these factors combined easily explain my newfound love for island bats.

While having to look for a needle in a geographically limited haystack, the value of other people's expertise beyond what's been published became apparent to me as well. There's a lot of information sitting in people's brains that hasn't made it into print or online, and that information is a lot more difficult to access because no search engine in the world will help you with it. This highlighted the value of networking and showed me how much more work I had to do to plan my trip if I wanted it to be successful. Luckily for me, I'd already started my blog at the time, and the post I'd written on the bats of Seychelles had already got me a bit of a reputation as a travelling Batman, one I'd inevitably strengthen during the BBY. That reputation helped me enormously when it came to getting in touch with people, as many had already heard of my project and were happy to help me by sharing some of the information about their local bats they'd collected over the years.

The trip ended on a rather sour note, as AirFrance/KLM decided to cancel my flight without offering me a seat on another plane. It took a few hours of calling various numbers that didn't seem to exist, a few unsuccessful calls to the Paris headquarters and finally, my parents getting in touch with the Belgian office for me to get on another flight (this would be the first of many issues with this airline during my bat-related travels). The only comforting thought in all this was that during my Big Bat Year, I'd have Stuart from Round the World Flights to assist me.

Tel Aviv, Israel, September 2019

I met Arjan and his colleagues to release a group of Egyptian Fruit Bats *Rousettus aegyptiacus*, after the end of a research project. After battling Tel Aviv traffic for a couple of hours, we were able to release the bats at sunset. Shortly after Arjan's colleagues had left, we spotted a Golden Jackal, a new mammal for me, right before we went on a short, ten-minute walk to a cave. This cave should have contained European Free-tailed Bats *Tadarida teniotis* and *R. aegyptiacus*. Unfortunately, all the bats appeared to be out foraging. At least the cave system was beautiful. On the recorder, we quickly identified Kuhl's Pipistrelle *Pipistrellus kuhlii* and *T. teniotis*, two common species in Israel and two I couldn't find in Belgium, so I was happy to bag them there. As we left the cave, we heard two rockets being fired from the Gaza Strip, followed by the response of the infamous Iron Dome. A few minutes later, a notification on Arjan's phone confirmed we hadn't imagined the whole thing! This was as close to batting in a war zone as I would get that year.

Our next stop was a pond south of the Dead Sea that Arjan knew well, as they'd done studies there using microphone arrays with the Geoffroy's Trident Leaf-nosed Bat *Asellia tridens*. It can be a tricky species to find as, like many desert species, they tend to occur over large areas but in relatively low densities. Knowing where they regularly forage is helpful. Squeezing through the reed bed, we came to an opening where we had to wait a while before an *Asellia* appeared, but eventually it came. As I didn't know of any roosts of that species, I'd figured this would be one of the most difficult ones to record, along with the Desert Long-eared Bat *Otonycteris hemprichii*. Arjan had planned for us to keep batting all night. However, I'd already been driving for a few hours, and more driving in the middle of the Negev desert didn't seem too appealing. Hence, we eventually went back to Tel Aviv, where I spent the night in his flat after we had the most expensive slice of pizza I'd ever seen, equivalent to about €15 for a single slice. That's the price of an entire quality pizza in Belgium.

As mentioned above, one species I was keen to find in this part of Israel was *O. hemprichii*, one of the few completely white bats in the world. Given my love

Habitat of the Desert Long-eared Bat *Otonycteris hemprichii*.

for long-eared bats, this seemed like a must-see. Again, being a desert bat, it tends to be widespread but rare. Arjan's colleagues suggested visiting the Ben Gurion Tomb, as one was known to forage there regularly. *Otonycteris* is probably one of my favourite bat species out there. I really love long-eared bats in general, as long as they're actually long-eared. I mean, I guess the Australian long-eared bats (*Nyctophilus* spp.) are long-eared, to some extent. They probably think they're long-eared, too, given they don't actually know the bats from the rest of the world that well. Social media isn't really in yet with bats. It would be interesting to see what they would share on Echobook. Would they constantly post about us like we constantly post about them? Or would they keep posting petitions to ask for our removal? Perhaps that's what's been holding them back.

Eilat, Israel, September 2019

The area around Eilat is famous among birders for the Champions of the Flyway event. It's an international competition where teams have 24 hours to see as many bird species as possible. Over the years, the money raised by the different teams has exceed half a million US dollars, all for bird conservation, which is incredible! I'd wanted to visit the area for a few years now, perhaps not necessarily during the competition – even though I think I'd like to take part one day. This place is especially suitable for such an event, because it allows easy access to a wide range of habitats, and it's located on a major migration route. Spring migration is more impressive because birds are in breeding plumage and singing and everything, but the autumn is still exciting. I figured that if the area was so good for birds, it would likely be equally suitable for bats too. I was especially drawn to the diversity of habitats. Having ponds, date plantations, rocky outcrops and desert areas all within short driving distance of my accommodation appealed to me. There were still a few common species I hadn't seen while out with Arjan. I would have liked to target the Eastern Barbastelle *Barbastella leucomelas*, but I wasn't able to find anyone who'd seen more than one or two in their entire career, suggesting it is a rare species. The targets I considered more reasonable were Christie's Long-eared Bat *Plecotus christii*, Botta's Serotine *Eptesicus bottae* and the Big-eyed Mouse-tailed Bat *Rhinopoma cystops*.

Eilat wasn't at all what I'd imagined; I was surprised to find myself in a beach-side resort type of city. Perhaps I could have figured it out – it's to be expected from a town on the Red Sea. Beyond identifying neighbouring countries in the distance such as Jordan and Saudi Arabia, I also tried to spot a few birds. I had two targets, the White-eyed Gull and the White-cheeked Tern. I only found the latter, on a buoy that I believe indicated the limit of the swimming area. It's weird to be walking on the beach with binoculars and pointing them at people, even when looking for birds, so I did my best to direct them away from everyone, which meant I kept scanning the buoys beyond. Luckily though, birds aren't huge fans of noisy swimmers either, and they stayed away from the people too. I also spent an outrageous amount of time trying to see exactly where the border with Jordan was along the coast. I'd seen it further inland – it was difficult to miss given the impressive fencing and checkpoint I'd passed – but it wasn't that obvious along the coast and in the sea.

The next day I decided to go birding at the famous K20 saltpans. As they don't seem to have an official name, the many birders visiting the area named them after their location relative to the motorway. This location is known to attract thousands of waders and gulls. While it's better to visit it in the spring, when many waders would be in their colourful summer plumage and species such as White-winged Black Terns would be very numerous, I figured it was worth a visit. The potential species that would be new for me were the Broad-billed Sandpiper and Armenian Gull. The former is a somewhat regular vagrant in Belgium. This was an excellent chance for me to finally catch up with the species.

I've always had a special love for waders. I met a few birders who told me there was one present, so I tried my best to find it. The problem was I didn't have a scope, and the light wasn't great, so many of the waders were backlit, making identification much more difficult. Scanning through the thousands of birds present was no small task, especially using just binoculars. Still, I took my time and eventually found a few Armenian Gulls, which I was happy about. After trying a few vantage points, I found a Broad-billed Sandpiper after about an hour or so. This was a good birding day for me. No major rarity, but I didn't care; I got a new wader. That was almost as good as getting a new bat!

Speaking of bats, I thought I'd come back in the evening, hoping this would be a relatively good place to find them. While the saltpans themselves wouldn't necessarily attract bats, there were date trees around as well as a couple of freshwater ponds. But unfortunately, bat activity was very low during the entire time I was there. Perhaps it picked up later in the evening; I don't know. This is a case where passive monitoring would truly shine, because it would allow for all-night recordings without screwing up one's sleep schedule. While most bats are usually out and about in the early evening, some bats can be more active late at night, and some could also be roosting quite far away, and their commute might take a while. The highlight of that evening was a Golden Jackal that spooked me while I was taking a break in the reeds. I initially thought it was a dog and could feel my adrenaline levels shoot up, but as soon as I realised it was a jackal, I knew I was fine. Coincidentally, that's also when the jackal thought it wasn't fine and ran off.

On the advice of Noam Weiss, the managing director at the Eilat International Birding and Research Centre, a well-known birder who happens to have done a bit of bat work too, I decided to visit an organic date plantation 30 km north of Eilat. He said it should have good levels of bat activity, giving me a decent chance at finding the species I needed. I wasn't going to be mist-netting there, so he told me I only needed permission from the people at the entrance of the property. When I arrived, there was a small group of people on the front terrace of the building. I stopped, and one came closer. He spoke perfect English, which made things much easier for me. He did seem a bit surprised by my request to drive around the plantation with a bat recorder but was happy to give me permission, nonetheless. He told me I should pay attention to what they called 'The Bat'. I wasn't clear on exactly what this was, but I understood that it was a machine used in the plantation. I spent a good two hours there, driving very slowly and stopping very regularly to listen to the bats. I was able to record *Eptesicus bottae* very early on. *Plecotus christii* took some more work, which isn't surprising given the quiet nature of this genus's calls, but I eventually got one flying close enough to the car that I was able to make recordings of it. I was also recording lots of Naked-rumped Tomb Bat, *Taphozous nudiventris* and Mouse-tailed bats (*Rhinopoma* spp.), including the Big-eyed Mouse-tailed Bat *R. cystops*, my last successful target for the evening and for Israel as a whole. This visit to an unexpected batting hotspot brought the year's tally up to 316 species. Over 270 days into this world tour and more than 300 bats later, I was holding my average of more than one new bat species per day.

As I was already pretty happy with the bats and birds I'd seen, I decided to keep exploring the area but without any pressure whatsoever to find anything, which is always the best kind of exploring to do. I did want to try to find a gazelle or two, though. I decided to visit Holland Park on the outskirts of Eilat. It's an interesting place with quite a few trees that seemed to have attracted a lot of migrants, mainly warblers of various kinds, but as I drove past those trees and got into the rockier part of it, it became full of wheatears. Wheatears are fun birds to watch, and Israel has a good range as well, making them exciting little ID challenges. They aren't particularly difficult to identify, especially for experienced birders. Yet my experience with them was limited, so I did want to spend some time with each one to make sure I was getting the identification right. In the end, I managed to find five species: Mourning, Isabelline, Hooded, White-crowned and Northern. I'd already seen Isabelline and Northern in Belgium, but three new wheatears was more than enough for me to be happy with my late afternoon/early evening outing. I did not manage to spot any gazelles or Sand Partridges but considering this was unlikely to be my last time in Eilat, it didn't really matter.

Following a mostly unsuccessful road trip to Spain and Portugal, where I was only able to add two species to my list in a little over a week, I booked a short trip to Tenerife, which I knew quite well, to add another two: the Madeiran Pipistrelle *Pipistrellus maderensis* and the Tenerife Long-eared Bat *Plecotus teneriffae*. Overall, I hadn't missed too many European species. I'd missed the Northern Bat *Eptesicus nilssonii* and the Particoloured Bat *Vespertilio murinus*, because travelling to Northern Europe felt too risky – I had no information on a reliable location for them in September (I learned much later than Latvia is great for them in the early autumn). There was also the Giant Noctule *Nyctalus lasiopterus*, which I'd missed in Seville. The issue I had in Europe was that I ended up losing a lot of time to get each species. This time could have been used more efficiently in other parts of the world. The reason I had so much time on my hands was that I was unable

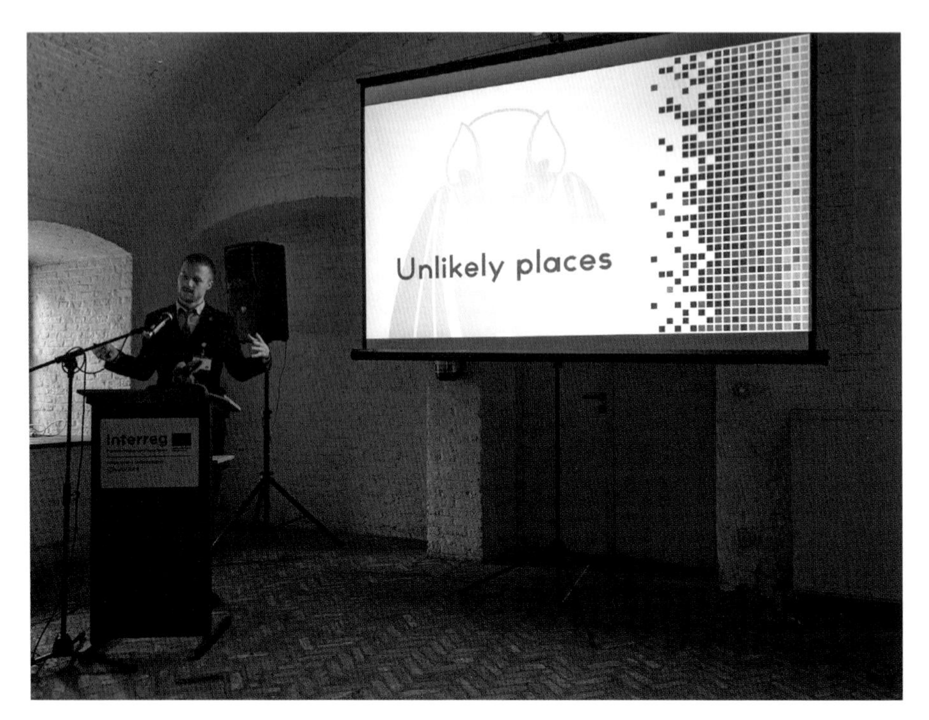

Conference on the Big Bat Year while it was ongoing.

to arrange anything in West Africa as I had intended, and I didn't have time to come up with alternative plans. Clearly, spending a week per species in Europe isn't the ideal way to do it – but it was fun to look for species I knew already, in places I was mostly familiar with. I also took advantage of my few remaining days in Belgium to speak at a conference on bat migration in Bruges. I'd started to really enjoy sharing my stories and my experience of travelling around the world to see bats. When I got the invitation to speak at that conference, I was quick to accept it even though all I could really think about was the last leg of my journey, in South America.

Bat migration

We've all heard of at least one migratory animal, probably even seen stunning images of this magnificent phenomenon on television. When you read the word 'migration', chances are the image of a Monarch Butterfly, a wildebeest or a tern popped into your head. The migration of Monarch Butterflies from the USA down to Mexico and then back up in a couple of generations creates seas of orange fluttery insects. The migration of millions of wildebeests in Africa creates storms of dust that can be seen from miles away. The migration of terns is equally impressive due to the sheer distance these relatively small birds cover every year. All those are incredible opportunities for some stunning images, made even better with a voice-over from our beloved Sir David Attenborough! Yet these are only a few examples of the many wonders of migration. Bats may not hold the record for the longest migration, the largest one, or any other related record for that matter, but that does not make their migration any less impressive.

One truly impressive bat species is capable of a transcontinental flight of over 2,000 km. At the time of writing, the record is over 2,224 km (that's 1,382 miles for our non-metric friends),[1] and it is held by an animal smaller than most objects you have around you right now. It is Nathusius's Pipistrelle *Pipistrellus nathusii*. To put this into perspective, that's three cubes of sugar flying from Latvia to Spain. It was named after Hermann von Nathusius, a German animal breeder who studied artificial selection at the same time as Charles Darwin, though he disagreed with Darwin's idea of evolution. In the autumn, many of these tiny bats migrate from north-eastern Europe to the south-west, looking for milder weather in which to spend the winter. The British population's relationship with the continent is as yet undetermined, but there are likely some exchanges, as some ringing records have shown (unpublished data).

However, the actual world record when it comes to bat migration distance is most likely held by the Straw-coloured Fruit Bat *Eidolon helvum*, a species weighing about 300 g – so much larger than Nathusius's Pipistrelle. It has been recorded travelling over 2,500 km from the Democratic Republic of Congo to Kasanka National Park in Zambia.[2] The exciting thing about this migration is that most individuals of that species from West or East Africa seem to converge on this national park between October and December.[3] The weight of this species allows for satellite tracking, something that isn't yet possible with the pipistrelle. Satellite tracking, unlike ringing records, gives much more accurate distance estimates and also provides information on the actual routes taken by the animals in both directions. This additional data is of paramount importance to any conservation plan, as the stop-over areas are likely as necessary as the origin and destination.[4] One record *E. helvum* certainly holds is being responsible for the largest mammal migration on earth. That's right – as impressive as the wildebeest migration is, these bats are even more incredible!

While these are two excellent examples of what bats can achieve, it's unlikely that they are a good representation of most bat species. This is especially true in tropical regions, as those species have few incentives to go on such journeys – there is plenty of food available for them year-round.[4]

Notes

1 Alcalde, J. T., Jiménez, M., Brila, I., Vintulis, V., Voigt, C.C. and Pētersons, G. (2020) Transcontinental 2200 km migration of a Nathusius' pipistrelle (*Pipistrellus nathusii*) across Europe. *Mammalia* 85 (2). https://doi.org/10.1515/mammalia-2020-0069

2 Richter, H.V. and Cumming, G.S. (2008) First application of satellite telemetry to track African straw-coloured fruit bat migration. *Journal of Zoology* 275 (2): 172–6. https://doi.org/10.1111/j.1469-7998.2008.00425.x

3 Ossa, G., Kramer-Schadt, S., Peel, A.J., Scharf, A.K. and Voigt, C.C. (2012) The movement ecology of the straw-colored fruit bat, *Eidolon helvum*, in Sub-Saharan Africa assessed by stable isotope ratios. *PLoS ONE* 7 (9): e45729. https://doi.org/10.1371/journal.pone.0045729

4 Fleming, T.H. (2019) Bat migration. In Jae Chun Choe (ed.) *Encyclopedia of Animal Behavior*, p. 605. London: Academic Press. https://doi.org/10.1016/B978-0-12-809633-8.20764-4

Puerto Maldonado, Peru, November 2019

I'd been invited by Chris Ketola, a Canadian wildlife fanatic living in Peru, to spend some time with Fauna Forever along the Tambopata River in the Amazon Basin. Before heading to the jungle, I spent a few days in Lima, hoping to find urban bats as I had in so many other cities. I was unable to find any and another attempt in 2021, in better parts of the capital city, also wasn't very successful. As far as urban batting goes, I wouldn't recommend Lima, even though local researchers have had far better luck than I did, albeit mostly in the outskirts. I also took this opportunity to go out at sea for some birdwatching. My main target was the Humboldt Penguin. I booked myself onto a fur seal cruise, which I was hoping would also offer me some proper seabirds. This venture ended up firmly in the 'successes' camp, with Peruvian Pelicans, Peruvian Boobies, Guanay Cormorants, Wilson's and Elliot's Storm-petrels and of course, Humboldt Penguins all coming into view during the cruise. What made the cruise even better was that the other people on the boat were keen to learn about the birds and kept asking me questions – including Jan Roberts, an American globetrotter who wanted to swim with fur seals, like most people on the boat. Jan was also very keen to quiz me about bats, as she was fascinated by my journey. Since then, she's become a keen bat champion, spreading her newfound love of them all around. Encounters such as this are incredibly motivating when on a journey like the Big Bat Year.

Chris had been stuck in Puerto Maldonado for weeks because he was recovering from a parasite infection. He made the best of his misery by building a triple-high net. Commercially available solutions for stacking three mist-nets on top of each other in order to target higher-flying species cost over two grand and are difficult to source in Peru. Chris was keen to see what he could build himself for a fraction of the price. My arrival brought the first opportunity for him to test it out and so we did. The location we chose was nothing special – it was a garden in the middle of town and yet, we caught a good selection of species. The first bats to get caught were *Molossus* species, the sort you'd really want to catch using a net like this, but they were quickly followed by Greater spear-nosed Bat *Phyllostomus hastatus*, the sort of species to destroy a net in a few seconds, cutting through them faster than the time it's taken you to read this paragraph. This encounter with one of the heaviest bats in the Neotropics was also my first encounter with a bat whose ferocious defence when caught could have been mistaken for the devil himself. Fruit-eating bats pack a powerful bite that they're not afraid to use when threatened by crazy bat researchers such as ourselves.

The garden was only fun for one evening, we needed to move on. Our first proper trapping surveys were at a site called Boca, where we were accompanied by a group of students on an expedition with the travel firm Where There Be Dragons. They had wide-ranging levels of interest; most of them weren't interested in what we were doing, in fact. However, a few students were really keen and joined us any time they could. It was nice not to just be trapping to find species for my list but also as a way to educate enthusiastic naturalists on how bat research works, what sort of features we're looking for once we've caught bats, and so forth. One of the keener students, who was really into reptiles and amphibians, spotted a snake high up in the vegetation. It took some effort to get it down, but when we did, we found it was a Brown Sipo *Chironius fuscus*. One could argue that we had unnecessarily disturbed this snake, and it's true we did so. However, while the students with us were the more nature-inclined of the bunch, most of them had never seen a snake up close before. Even I hadn't examined one this close, and that in itself is a learning experience. Chris gave us a brief explanation of what's needed to safely handle these reptiles, so we all left with some more valuable knowledge and the snake went on its way unharmed. Chris doesn't get too many encounters with snakes, and that is why he catches, measures, identifies and sexes every snake he sees. When you have so little data, every additional entry counts.

As it was my first time in South America, new species of bats were coming in quickly, reaching a satisfactory eight species by the time we headed back to Puerto Maldonado, where we added another five on the bat recorder. Most of the species I was finding were common, nothing Chris really got excited about, but one of the first bats we caught at Boca was the Fringe-lipped Bat *Trachops cirrhosus*. This phyllostomid is well known for specialising in frog-hunting. It's capable of recognising the calls of species that aren't toxic. Studies initiated by Merlin Tuttle and Mike Ryan have shown that these bats quickly learn new calls and that they have had major impact on how frogs court. Subsequent research by Rachel Page has even demonstrated that bats trained, marked and released can still remember unique calls they were taught while in captivity years earlier. A truly fascinating species.

The boat ride to Explorer's Inn, our base for the remainder of my stay, took nearly four hours. We stopped at a few clay-licks, places where clay is exposed and which attract a wide range of animals in need of a mineral fix. While the main highlight is the colourful macaws, clay-licks attract far more than that. On the mammal side, we spotted a family of Capybaras on our first stop, the largest rodent on earth. Unfortunately, a couple of other mammals we had no chance of spotting were manatees and dolphins. The Tambopata River, a tributary of the Amazon, has been dammed downstream of Puerto Maldonado – meaning that all large animals are now unable to swim up it. I was surprised to see how murky the waters were; it was hard to imagine anything living in them at all. The water wasn't always this colour – it's been made worse by the intensive gold-mining along the banks. It may be illegal but that doesn't make it a rare occurrence.

Unsurprisingly, so deep in the jungle, even around camp, we were able to find new species for my list, including the Brown Mastiff Bat *Promops nasutus*, with its typical upward curving calls, similar to the ones I'd recorded on my last day in Yucatán. The Greater Sac-winged Bat *Saccopteryx bilineata* as always was the most commonly recorded species, as it appeared to be absolutely everywhere. Luckily, it's easy to identify based on echolocation and doesn't really overlap with anything else, making it easy to ignore when analysing call sequences. We took full advantage of the fact that we now had three static recorders to scout potential trapping areas, assessing their bat activity and diversity. One species I'd missed in Mexico, the Proboscis Bat *Rhynconycteris naso*, turned out to be quite regular on those recordings but I couldn't count it. I'd excluded static recorders from the ways I could add species to the Big Bat Year list, and I'm not one to break my own rules. We were hoping to catch it at some point, or even find it roosting somewhere, but that didn't happen. I ended up having to leave Peru without one of the bat species most commonly seen by people on tours in the Amazon Basin. My frustration was immeasurable.

We'd planned to mist-net close to an oxbow, a lake formed by a river interrupted on either side. On the way there, we heard rustling in the leaves on the forest floor. I got excited because I thought it was a mammal of some sort. That excitement disappeared quickly when we realised it couldn't be a mammal because it wasn't running away, which is the one thing all mammals have in common – they don't like humans. OK, there are exceptions, but not that many. If it wasn't running away and it was moving rather slowly, on the ground, there's only one thing it could be: a tortoise. Specifically, it was a Yellow-footed Tortoise out for a stroll. As I'd never seen a wild tortoise before, my levels of excitement went right back up. After we took some photos of it, we headed to the lake. We started by setting up a bat recorder over the water, attached loosely to a fallen tree, meaning no branches or foliage were obstructing the recorder, which is crucial to get decent-quality recordings. We didn't know what species we could expect on that lake, but it was worth a shot.

Usually, open places, especially water, attract a range of bat species. The open surroundings help with the acoustic identification; it can be rather tricky in cluttered environments, as is the case inside the forest. After placing the recorder, we opened a few nets, just three because we didn't want this to be a late night. It's

A smoky Puerto Maldonado.

always tempting to open lots of nets. However, later in the evening is when you regret that decision – having to check them all, keep up with all the bats that get caught and at the end of the night, put all the kit back in the backpacks! The bat activity was surprisingly quiet. We only caught a handful, most of them Seba's Short-tailed Bat *Carollia perspicillata*. We had to move our processing station a couple of times because leafcutter ants seemed interested in our backpacks and datasheets. They'd already managed to cut an almost perfectly circular hole in my backpack by the time we'd realised. Admittedly, it wasn't a very sturdy one, but still, they're incredibly efficient at cutting things. One of my AudioMoths, in addition to yet more sequences of *R. naso*, recorded the Lesser Bulldog Bat *Noctilio albiventris*, a species both Chris and I were very keen to catch. The last surprise of the evening came in the form of baby caimans calling for their mum when we went to pick up the recorder before heading back to camp.

Over the many hours of trapping, I discovered how intense Phyllostomidae could be. I'd had an introduction to them in Mexico, but I hadn't handled any before coming to Peru. They are very keen to show that they're not to be messed with. Despite being fruit eaters, they are extremely well equipped to defend themselves, possibly to fend off predators – but this also makes their handling more challenging. I can get quite stressed while handling bats, particularly if I think I'm doing it wrong, which Phyllostomidae often gave me the impression of; as a result, I started noticing that I was enjoying our trapping sessions less and

Tambopata River.

less. Coincidentally, as the nights flew by, we were catching increasingly fewer new species for me. It was hard to stay motivated to put in the effort to hike for a while, deploy the nets, extract the bats and process them, when most of them were bats we'd already caught dozens of times. But that's the reality of being a bat researcher. It just wasn't a reality I could handle while running the marathon of my Big Bat Year.

My tiredness grew to such an extent that I didn't feel comfortable safely heading out in the evening for another trapping session, so we skipped a night. Luckily, earlier that day, Chris and I had walked some of the trails in the search of possible tree roosts. We ended up spotting the Buffy Broad-nosed Bat *Platyrrhinus infuscus*, a new species and one I wouldn't see later. Worth the effort of wrapping a net around a tree while making sure all the exits were blocked. There was definitely lots of creativity involved in this capture. Chris convinced me that we should take a day off and then press on the next day to trap at another oxbow lake, one that looked extremely promising on paper and one Chris had never trapped at. The lake was more than an hour's walk away from camp. Laurie, an intern on the bird team, was keen for us to do some birding while we were there. As this sounded like a good idea, we left early to ensure we'd have plenty of time to enjoy the birds before it was time to set up the nets. This turned to be an excellent plan as we saw a few nice birds, including the Green Ibis, Western Striolated Puffbird and Green Kingfisher. We also found caiman tracks with the typical claw and tail marks, characteristic of all crocodilian tracks. Sadly, the caimans themselves were nowhere to be seen.

A Spectacled Owl kept us company while we were putting up the mist-nets. Once the nets were up, the party started – quite literally for some, as we could hear loud music playing from the strictly protected forest. Bat activity above the lake was incredible but most of them were too far from the banks to catch. However, one cleverly placed net turned out to be the golden net of the evening; it caught four *N. albiventris* exiting the forest. All bats were caught on the forest side of the net that we'd placed along the bank, and that led us to believe that we'd put it right on the bats' flight path as they went foraging on the lake. Both bulldog species are highly specialised fishing bats, but they need hollow trees or caves in which to roost. We were incredibly lucky to catch them without resorting to putting up nets in the middle of a pond, which comes with serious logistical problems and safety concerns.

That night, I became acquainted with the deadliest living organisms in the rainforest... trees. Probably not what most people would expect, but bushmasters, wandering spiders and caimans injure and kill far fewer people than trees do when falling down. Unlike most trees in temperate regions, the root systems in the rainforest are often very shallow, giving the tree very little resistance against being swayed sideways. While walking in the forest, it's far from unusual to hear one fall down. When that happens fewer than 100 m away from you, and it

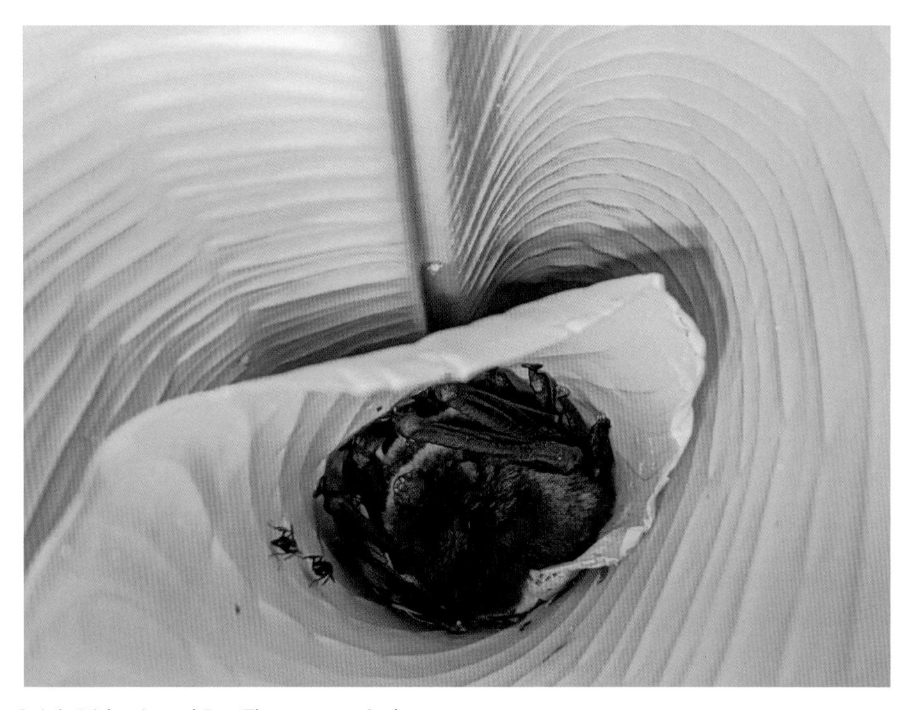

Spix's Disk-winged Bat *Thyroptera tricolor*.

happens to be a 40 m tree, it is pretty scary. And when it happens to fall on the trail you're supposed to take, making it vanish entirely into the forest, it's even more frightening. That is precisely what happened to us that night. 'Had it not been for Chris's experience dealing with such things, we'd probably have got hurt quite quickly. He told Laurie and me to stay put while he went to search for the trail with his machete, torch and most importantly, handheld GPS with the trails mapped on it. We eventually got back to camp, safe and sound.

My thirtieth new species in Peru was the Dwarf Little Fruit Bat *Rhinophylla pumilio*, quite a rare one that Chris hadn't caught all that often. It took us a while to get to an identification we felt confident in. Phyllostomidae can be extremely tricky sometimes, especially when bats start to pile up in the net and pressure rises to release the ones already caught. Clearly, I could have found more species, had we not skipped a night or two of trapping – but then we'd pretty much run out of common species to find. Rarer species were exponentially more effort and you had to get past all the *Phyllostomus* and *Carollias* to get there, which I wasn't so keen to do. Trapping in the Neotropics isn't my thing.

Food habits

Bats have some of the most diverse diets of all mammalian orders. One family in particular, Phyllostomidae, shows an astonishing diversity of feeding habits. Phyllostomidae eat everything other bats do and more.

Most people probably think of insects as the primary diet for bats, and that's fair given that over 60% of all bat species consume them. When it comes to lesser-known diets, that's where this diversity truly becomes interesting. The second most common diet is fruit-based. Most of the time, only the fruit pulp is valuable for the bats themselves; the seeds are then excreted. That is why bats are key actors in the regeneration of forests, just like birds. They've even been shown to complement each other because they feed on different species of trees and shrubs, albeit with some overlap.[1,2]

In the New World, 145 genera and 191 species of fruiting plants are consumed by frugivorous bat species. Nectar is another common source of food for a number of bat species, especially from the Pteropodidae and Phyllostomidae families. Similarly to hummingbirds, a number of bat species have evolved specific features to make the task of getting to the nectar easier, such as extremely long tongues (for example *Anoura* species can have tongues one and a half times the length of their own body[3]), long snouts such as those seen on *Agoura*, *Macroglossus*, *Megaloglossus*, and so on. Fruit- and nectar-feeding bats occasionally munch on leaves, blossoms and insects to supplement their diet.

One bat food that's surprisingly well known is blood. Well, I say surprising – it isn't surprising that this diet is ghoulishly interesting to most people. It is surprising, however, how commonly it's spoken of, given only three species (that's 0.002% of all bats) feed on blood.

The more obscure diets are also the more interesting ones, in some respect. A dozen or so species have a primarily carnivorous diet, meaning they'll feed predominantly on small mammals (including other bats), small reptiles, amphibians and fish. Some interesting evolutionary traits lead certain species to be more effective at catching such prey. The Fringe-lipped Bat *Trachops cirrhosus* is known to hunt frogs by listening to the males' mating calls. This bat is also able to distinguish frog species based on their calls and exhibits a different response based on call type.[4] *Trachops* isn't the only predator bat to hunt by listening for potential prey; other species such as the Pallid Bat *Antrozous pallidus* and the Heart-nosed Bat *Cardioderma cor* have also been shown to do so,[5] but perhaps not with the level of complexity that comes with *Trachops'* interpretation of the different frog call types. In fact, two bat species have been shown to not react to frog and cricket calls, only to their rustling on the ground.[6]

Fishing isn't something one would commonly associate with bats. Yet every continent has one or more highly specialised bats that make them more effective at catching prey on or below the water surface. The most common and perhaps also most obvious adaption is the large feet which most trawling species exhibit, especially in the *Myotis* genus – *M. capaccini, M. vivesi, M. ricketti* and *M. Macropus*, for example.[7] A few non-*Myotis* have evolved huge feet too, such as *N. albiventris* and *N. leporinus*. The latter has developed the capacity to use echolocation to detect ripples on the water's surface, which it can then interpret as fish movement underneath.[8] This 'indirect' use of echolocation is quite remarkable!

Notes

1 Ingle, N.R. (2003) Seed dispersal by wind, birds, and bats between Philippine montane rainforest and successional vegetation. *Oecologia* 134 (2): 251–61. https://doi.org/10.1007/s00442-002-1081-7

2 Medellin, R.A. and Gaona, O. (1999) Seed dispersal by bats and birds in forest and disturbed habitats of Chiapas, Mexico. *Biotropica* 31 (3): 478–85. https://doi.org/10.1111/j.1744-7429.1999.tb00390.x

3 Muchhala, N. (2006) Nectar bat stows huge tongue in its rib cage. *Nature* 444 (7120): 701–2. https://doi.org/10.1038/444701a

4 Jones, P.L., Divoll, T.J., Dixon, M.M., Aparicio, D., Cohen, G., Mueller, U.G., Ryan, M.J. and Page, R.A. (2020) Sensory ecology of the frog-eating bat, *Trachops cirrhosus*, from DNA metabarcoding and behavior. *Behavioral Ecology* 31 (6): 1420–8. https://doi.org/10.1093/beheco/araa100

5 Bell, G.P. (1982) Behavioral and ecological aspects of gleaning by a desert insectivorous bat *Antrozous pallidus* (Chiroptera: Vespertilionidae). *Behavioral Ecology and Sociobiology* 10 (3): 217–23. https://doi.org/10.1007/BF00299688. Ryan, M.J. and Tuttle, M.D. (1987) The role of prey-generated sounds, vision, and echolocation in prey localization by the African bat *Cardioderma cor* (Megadermatidae). *Journal of Comparative Physiology A* 161 (1): 59–66. https://doi.org/10.1007/BF00609455

6 Jones, P.L., Page, R.A. and Ratcliffe, J.M. (2016) To scream or to listen? Prey detection and discrimination in animal-eating bats. In M. Brock Fenton, Alan D. Grinnell, Arthur N. Popper and Richard R. Fay (eds) *Bat Bioacoustics*, pp. 93–116. New York: Springer. https://doi.org/10.1007/978-1-4939-3527-7_4

7 Aizpurua, O. and Alberdi, A. (2018) Ecology and evolutionary biology of fishing bats. *Mammal Review* 48 (4): 284–97. https://doi.org/10.1111/mam.12136
8 Schnitzler, H.U., Kalko, E.K., Kaipf, I. and Grinnell, A.D. (1994) Fishing and echolocation behavior of the greater bulldog bat, *Noctilio leporinus*, in the field. *Behavioral Ecology and Sociobiology* 35 (5): 327–45. https://doi.org/10.1007/BF00184422

Galápagos, Ecuador, December 2019

Only two bat species roam the skies of the Galápagos, and my intention was to spend four days there. How could that possibly make sense? The short answer is that it didn't. The longer answer is that it was the opportunity to chase my second lifelong dream. When I first saw the film *Master and Commander* in 2004, the scenes showing the islands truly fascinated me, primarily because of their unique wildlife. I knew I wanted to try to visit one day. Joining a cruise is expensive and well beyond my budget for the Big Bat Year. If I wanted to visit the islands there and then, I had to come up with a solid plan that would make it worth it both for my Big Bat Year and for me. Without a cruise, a few parts of the archipelago, and therefore many species, were out of bounds – chiefly, the Flightless Cormorant. While it is a remarkable example of evolution taking things in a bizarre direction, there are plenty of other animals, including kiwi, of which I'd seen three species. Hence, I wasn't that disappointed I wouldn't be able to see it. The rest, including the endemic penguin, could be seen by spending some time on Santa Cruz, Isabela and San Cristóbal. The capital of Santa Cruz, Puerto Ayora, has several places offering relatively cheap accommodation. The ferries to the other two islands are reasonably priced. They could not only take me to said islands but also give me opportunities to spot some of my favourite birds out there: seabirds.

I'd managed to put together an itinerary that wouldn't cost me an arm and a leg, but I'd underestimated the price of flights, visas, permits and whatnot. Those quickly inflated the final bill for this quick detour. But I decided they were worth it, and I had to go for it. So I did. Upon landing, it was quickly evident how easy some of the birds were to find, as there were already a couple of species of the so-called Darwin finches flying around the tarmac. I was expecting strict biosecurity checks, as they have in New Zealand (and Australia), especially given the US$100 entrance fee (that would be on top of the U$20 Galápagos visa). As we got inside the building, I was quickly disabused of that notion as the checks were only superficial and similar in their effectiveness to biosecurity checks in Belgium – that is, highly ineffective, I'd say. This made me slightly angry because of all the fuss that's made around the importance of protecting the islands, especially inside the airport building – but then there are no checks, no declarations that one must sign to support the efficacy of the measures. The theory did not live up to the reality. Well, I'm not being completely honest here; there were 'checks'. There was a sign saying we had to wait for the all-clear before collecting our hold luggage, because they had special dogs at work. I did see one of those dogs, and he was as helpful as the dog that brought JFK airport to a halt for some belly scratches, on my way back from Mexico. They most certainly did not investigate the luggage thoroughly. And then we were given the all-clear. Go figure...

Shortly after going through the fake biosecurity checks, I hopped on the first bus to the ferry terminal. In what I can only assume is an attempt to minimise the damage on Santa Cruz, the airport was constructed on a very dry islet just off it, Isla Baltra. Most people had arranged private transport, as I'd have expected, but many other people, especially young travellers, seemed to be taking the same approach as me, so I knew I was going the right way. At the ferry terminal, the proximity of wildlife hit me; I was spotting Galápagos Shearwaters and Blue Boobies from the pier. How crazy is that? Shearwaters at 30 m from the pier, I'd never seen that. I was used to having to find a boat to stay on for the day, in order to see those.

My frustration at the failures of the airport authorities quickly faded away as my enjoyment and my younger self's joy came back to make the most of this short trip. I spent the whole crossing photographing the birds. After hopping onto another bus for a much longer drive, and a brief stop on the road because of a traffic jam caused by a Giant Tortoise, I checked into my hotel and started figuring out how to book the ferries the following days. I'd seen on the website observation.org and in other trip reports that people had found both of the Galápagos bat species in the vicinity of the harbour, so that was my plan for the evening: have a stroll around town, find something to eat, then hang around the port looking for bats. If I were to critically evaluate how often plans like that actually work, it wouldn't be a very positive review.

Eleven months into my world tour, I can tell you that the biggest undertaking that hadn't gone according to plan was the Big Bat Year itself. Big Years are as much about getting back on your feet after setbacks as they are about planning. But this time was different; this time, my plan was perfect. After a surprisingly cheap seafood meal that included a lobster (the first time I could ever afford to buy myself a lobster without running out of money for the rest of the month), I went back to the harbour, and looked at some sharks hunting under the streetlights. (Are they still considered streetlights if 80% of their beam illuminates the ocean? I don't know.) The sharks didn't seem to mind being illuminated, and neither did their pelican friends. I'm not entirely sure what the sharks were, but I'm thinking something along the lines of

White-tipped Reef Shark. I'm not a shark expert, though, and I have to admit I didn't look it up at the time.

My bat recorder had been on for most of the evening, except for the lobster episode. It had been recording the local Hoary Bat *Lasiurus/Aorestes villosissimus* most of the time. Again, I'll spare you the 'Which genus is most correct?' discussion. The second species, the Galápagos Yellow Bat *Lasiurus brachyotis*, took some more patience. I can't say it was more work because all I had to do was stand watching the sharks for a little longer. Once I'd bagged those two species, I went for a bit of a stroll further along the shore to see what the rest of the town looked like. It was apparent that this town survives through tourism. Every other shop was a souvenir shop, and the ones in between were tour operators. I only had a vague interest in the first and none in the latter, so I walked past them. Once again, the proximity of wildlife was staggering; Marine Iguanas were everywhere early in the evening and could be approached easily for some close-up shots. A young Lava Gull I encountered didn't seem to mind the proximity to people either.

After a good night sleep, I was on the ferry to San Cristóbal. I'd been able to get tickets straight from my hotel's reception, which made the whole process very easy. My targets there were the seabirds I could see on the way, such as the Waved Albatross and Galápagos Petrel. On the island, I could see a couple of new finch species as well as the San Cristóbal Mockingbird. The ferry crossing was choppy – we were sailing across a stretch of the Pacific (admittedly a short one) – but I was still able to spot some great seabirds that made me unbelievably happy, somewhat to the bewilderment of the other passengers. The Red-footed Booby, Waved Albatross, Galápagos Petrel, Pink-footed Shearwater, Sooty Shearwater and Wedge-rumped Storm-petrel all came within sight during this relatively short crossing. I even saw a Red-necked Phalarope, species number 2,202 for the year. I wasn't doing a Big Bird Year but I wasn't doing too badly with birds either – I'd seen a little over 2,200 bird species so far compared with 363 bat species. At the start of the journey, I hadn't been picturing these kinds of numbers at all.

On San Cristóbal, I quickly got distracted by the colonies of sea lions lying on the various beaches, benches and piers I walked past. Once I made it past them, I could look at the birds and try my best to locate the three species I wanted to find: the Vegetarian Finch, Woodpecker Finch and Small Tree-finch. They can be tricky to ID because the differences in bill shape can sometimes be rather subtle. It took me a few attempts before finding individuals of each species that I was happy to confidently identify. The only species I didn't struggle to ID was the stunning Gálapagos Flycatcher. New World flycatchers had driven me a bit mad in Peru, but the diversity was limited enough here to be sure of my identification. The individual I found was also very inquisitive, so getting close views wasn't exactly a challenge.

After a short break in a restaurant with good Wi-Fi to catch up with my parents, it was already time to go back to Puerto Ayora – sadly, without seeing the mockingbird. The next day, the ferry journey to Isla Isabela had fewer birds but one species I hadn't seen in Ecuador yet, Elliott's Storm-petrel. Even though this species wasn't new for me, storm-petrels have held a special place in my heart ever since I worked on them during my MRes degree. However, I did see a Red-billed Tropicbird around the harbour, which was new for me. What I'd come to Isabela for, I hadn't seen yet, though. That was the Galápagos Penguin. Penguins fascinate me, as they fascinate many other people out there. By that point, I'd seen three species in New Zealand. I was keen to see the one species that's known to sometimes dwell in the Northern Hemisphere, where its unrelated Alcidae friends cause a lot of confusion in many languages, including French. In French, the word for 'auk', the colloquial word used to talk about members of the Alcidae family, for example puffins, guillemots and razorbills, is '*pingouin*'. You can see how most people would get confused. Away from the colonies, on the far side of the island, the penguins can be tricky to find, but it wasn't impossible to locate an individual swimming near the harbour, from what I'd heard.

There weren't any new finches, or any other bird that I could find that would be new, so I wasn't sure what I should be looking for. I found a pond with some shorebirds and waterbirds, which was fun. I hadn't had a chance to spot many of those in the Neotropics. I scored White-cheeked Pintails and American Moorhens. Sure, they're common species, but I hadn't seen them yet, so it made no difference to me whatsoever how rare they are. On the beach, a little bit further along, I spent a good couple of hours photographing Marine Iguanas, as the scenery around them was far more photogenic than in Puerto Ayora. I was distracted from my reptile photo shoot by two much more active reptiles, also known as birds – specifically, a Semipalmated Plover and a Least Sandpiper, both of which I tried to photograph. Overall, my short stay on Isabela turned into more of a beach photo session than a penguin search trip, but I was happy nonetheless. On the way back to Santa Cruz, I kept an eye out for possible penguins and did see one feeding near the rocks 200 m or so outside the harbour.

My Galápagos trip was a resounding success, one I will remember for a lifetime. Those four days will remain some of the best I had that particular year, and I'd had a lot of great days. This might surprise you given I only managed a quarter of my daily bat target, scoring only two in four days, but I'd made a second childhood dream come true, after discovering a new bat species in West Papua. I already had the world record for the number of bat species seen in one calendar year; I was allowed to have some fun along the way as well!

Creatures from another world.

Back on the continent, I checked into a hotel ready to be picked up early the next day. The hotel lobby had a very purple Christmas tree that I had to take a photo of to send to my dad, since the football team he supports; Royal Sporting Club Anderlecht, wears those colours. At six in the morning, the driver from Bellavista Cloud Forest Lodge was there to pick me up and on we went for a three-hour drive. At Bellavista, I found myself in a slightly uncomfortably tranquil lodge. My mind was swiftly put at ease by the arrival of Richard, the owner of the lodge, whom I'd met at the British Birdfair. He was happy to show me around the premises, telling me that I was essentially free to do whatever I wanted to study the bats in the area. While not quite as exciting as the batting prospects, the birding prospects from my bed were still more than desirable to a not-so-early-bird type of birder like me, because I had a 360-degree view from the top of the wooden tower I was sleeping in. One issue remained, the weather; it was pretty much raining constantly, not a heavy downpour but a continuous light rain that made everything slightly more difficult. I spent my days walking around the lodge, talking to Martin and Amanda, a Welsh volunteer couple, about my travels and theirs, helping clear out trails and looking at the recordings from the night before.

I was moving my AudioMoths from one place to another to get a sense of where there was good bat activity. I skipped trapping for the first two nights because of the rain. However, it had become obvious that I couldn't keep on waiting for dry evenings because those were unlikely at this time of year. Given I was still getting a fair bit of activity on the recorders, I figured I might as well try and deploy three nets, one near the hummingbird feeders and two on nearby trails. I had no intention of walking 2 km alone, in the dark, through the cloud forest, so I found spots very close to the lodge. Trapping on my own still wasn't something I enjoyed all that much. But I also knew that it was my only way to see species that don't echolocate often, such as the majority of Phyllostomidae, and cloud forests have a unique diversity of such species, especially in the *Anoura* and *Sturnira* genera. *Anoura* spp. are regularly spotted in high-altitude birding lodges that have hummingbird feeders, mostly because they're a bit like the hummingbirds of the night. Like them, they feed primarily on nectar, but any beverage full of sugar that's offered to them is welcome too. They have a long snout, not dissimilar to a hummingbird's long beak, and like them, they have a ridiculously long tongue. Placing nets right next to the feeders seemed like the best way to catch one.

Things rarely work as planned when it comes to bat trapping and the most productive nets are often the ones that are just setup randomly. And indeed, I caught Handley's Tailless Bat *Anoura cultrata* on one of the trails. That was the only bat I caught that night, or at all at Bellavista, but it was a good one. This *Anoura* is recognisable by the presence of a very short tail and a very short forearm. The other species I could have expected, the Peruvian Tailless Bat *Anoura peruana*, has no tail at all and a forearm about 25% longer. As my nets had attracted the curiosity of a couple of birders with their guide, I spent some time explaining how I'd caught it, how I identified it; they were quick to notice the similarities with hummingbirds, including the tongue sticking out.

The rain came back, and I closed the nets. They were wet and full of vegetation as I collected them. Nets full of vegetation can't be put away safely, as they'll rip the next time someone tries to open them. They have to be thoroughly dried and cleaned before being stored. This painful process took me most of the next day and that's when I decided I was unlikely to be trapping there again. It meant I had to give everything I could on the recorder. During my various walks around the lodge and on the road, I recorded some nice species such as Small Big-eared Brown Bat *Histiotus montanus*. One species was remarkable by its absence: Little Black Serotine *Eptesicus andinus*, which I assumed would be common, as *Eptesicus* usually are. I would have liked to catch *H. montanus* because it's a big-eared bat, and I'd seen one on each continent – *Plecotus* in Europe, *Nyctophilus* in Australia, *Otonycteris* in Asia and so on – but I hadn't seen a large-eared bat in the Neotropics yet. They are some of my favourite bat species out there, so I was looking forward to that. *Histiotus* is a mostly Andean genus, similar to *Plecotus* but more closely related to *Eptesicus*. Some of them, such as Tropical Big-eared Brown Bat *Histiotus velatus*, were originally described as members of the genus *Plecotus*. Like many bats around the world, others were assigned to the genus *Vespertilio*, which seemed very popular in the 1800s. Nowadays, *Plecotus* is a temperate genus. In tropical Asia for example, it's only found in the Himalayas, and it's almost completely absent from Africa with the notable exceptions of the high Ethiopian plateaux and Cabo Verde. The molossid activity was very high most nights, with little regard for the weather conditions. They can be challenging to identify, but the ones I could distinguish weren't anything special, despite the altitude. However, I did manage to identify Peale's Free-tailed Bat *Nyctinomops aurispinosus*, which was new for me. What surprised me the most about that sighting was that it was done under pouring rain. It hadn't stopped raining all day and the bats must have decided they had to feed, even in these suboptimal conditions.

Any time in the evening when I wasn't out looking for bats, I was waiting near the bananas that staff put out every evening to draw the local wildlife; it's become a top-rated attraction at the lodge over the years. Tayra, Andean White-eared Opossums, and squirrels all were regular visitors. One visitor was more special than the others: the Olinguito, a species discovered at this very lodge. The story of its discovery is a rather funny one; staff kept noticing bananas disappearing from their pantry. One day they saw the culprit, an Olingo lookalike. At the same time, researchers at the Smithsonian Institute were describing the newest and smallest member of the Raccoon family, the Olinguito *Bassaricyon neblina*. When it became clear this species was a cloud forest specialist, the identity of the Bellavista Lodge thief became clear! Eight years after its discovery in 2013, Bellavista is still one of the very best places to see the species. Evidently, its love for bananas hadn't disappeared because on the third evening, while out trapping, staff called me to tell me an Olinguito had come and was likely to come back. They knew I was keen to see one, so if I wasn't on the lookout myself, they would try to find me around the lodge to let me know when there was one present.

One of the excursions Bellavista offers, being a birding lodge, is a trip to a Cock-of-the-rock lek. Leks are areas where males of a particular species are known to gather to convince partners to mate with them. They are very popular with photographers because birds often display fantastic behaviours. Because many leks tend to remain for many years, it's possible to build hides to make it easier to photograph those behaviours. It's a popular bird photography attraction for Black Grouse, Greater Sage-grouse and, of course, Cock-of-the-Rock (COTR). The latter, despite being a very colourful bird, is not easy to spot outside of those known sites, which tends to be the most reliable way to see this species for anyone in the Andes. Doing this excursion on my own would have been expensive, so I tried to convince Martin and Amanda to come with me. Like me, they had a limited budget for all this, and because they were volunteering, they had work obligations. Despite this, they kindly agreed to join me on the COTR trip and the Oilbird trip the next day. Leks aren't restricted to birds. In fact, a few bat species are known to use a similar mating strategy, particularly in the Neotropics – such as the ugly Wrinkle-faced Bat *Centurio senex* (don't trust anyone who tells you they're cute).

Every cloud forest was unique and I loved all of them.

I woke up in the middle of the night, with the feeling the tower was moving, that everything was moving. I figured I was probably still asleep and didn't react in any way. Perhaps not the wisest thing to do, as I discovered the next day while talking with the staff that there had been a 4.5-magnitude earthquake, far bigger than the earthquake I'd once felt in Belgium that scored a 1 on the Richter scale. Luckily, there was no damages and no one was hurt, at least in that part of the country, but it could have been a lot worse, especially given my sleeping brain didn't seem to comprehend danger. What started as a worrying day, though, quickly turned into an exciting one; I was going to see an echolocating bird, the Oilbird. I'd seen a bunch of swiftlet species already, the other group of echolocating birds, but I really wanted to see an Oilbird. The cave we visited wasn't so much a cave as a canyon of sorts. The birds were perched on either side, surprisingly discreet at first. They were oddly elusive and until we'd seen the first one, they all seemed invisible. After a short while, however, we saw about a dozen or so. There was one small detail that helped us detect their presence: the noise. Oilbirds are obnoxiously loud. Friends of mine had told me stories of how loud one had been when caught in a bat net, and imagining the noise of an entire roost was a terrifying thought.

Antonio, a young chap of 65 years who was far more agile on his feet than me, after showing us the Oilbirds and going down a couple of culverts to find bats, was still keen to hike up a mountain slope to show us Cock-of-the-Rock. He said it was a 20-minute walk; I looked at the mountain, had flashbacks of my 'one-hour' hike in the Solomon Islands, explained said flashback to Martin and Amanda, and we agreed to call it a day – especially given we'd already seen them the day before. On the way back, I spotted a few more new birds, in habitats of mid-elevation, and was able to show some to Martin and Amanda as well, which made us all very happy. I'd already reached the end of my stay at Bellavista, and I was getting picked up the next day to go to Guango, another famous lodge in the Ecuadorian Andes, but located on the eastern slope.

Climbing up the eastern side of the Andes, I noticed that the landscape was very different. There was less cloud forest and more elfin forest, despite the altitudes being comparable. Elfin forests consist mainly of shrubs and bushes, and bamboo. It's the preferred habitat of the Spectacled Bear, among other animals. The road back up the Andes, on the other side, was very different; the landscape here was unique, which comforted me in thinking I'd made the right decision. I'd made plans to go on a guided birding tour of the Páramo – a place akin to the tundra of the far Northern Hemisphere, so I knew it'd be good either way, bats or no bats. I was also quite hopeful I'd get to see a Spectacled Bear as they are regularly spotted in the area, and the signs on the road highlighting their presence are numerous too. But that's not always an indication of anything. There are substantial differences in the bird species on the west and east side of the Andes; among bats, seemingly less so, but as I'd only caught *Anoura cultrata*, anything else would be new anyway.

Like at Bellavista, I was given permission to mist-net anywhere I wanted. I was able to communicate with one of the staff members thanks to the three whole sentences I'd learned in Spanish. I asked him to source some poles from the forest, which turned out to be difficult to use because they were heavy – but they worked fine all the same. Mist-netting poles can really be made out of anything. I didn't have much success trapping around the hummingbird feeders while at Bellavista, but the hummingbird garden at Guango was much larger. There were lots of new hummingbird species I hadn't seen yet, including the Sword-billed Hummingbird with its humongous bill. I put out two nets in the evening in between all the feeders and one across a forest trail not too far away. I'd found the promising location while out on a birdwatching walk where the guide showed me an Andean Potoo, yet another bizarre-looking bird in the nightjar family. I was happy to catch only three bats. I didn't want all my nets to be filled, as that would have likely caused injuries. Low bat activity when I'm out trapping on my own is not an issue, as it means I have ample time to look at each bat I've caught without being overwhelmed. .Every bat made sure to wait about half an hour after the previous one before getting caught. How civil of them! The first two bats were the Hairy Yellow-shouldered Bat *Sturnira erythromus*. The Andes have an impressive diversity of yellow-shouldered bat species,

Someone's been feeding on nectar, Peruvian Tailless Bat *Anoura peruana*.

far more than I had encountered down in the Amazonian lowlands. Therefore, the two individuals I caught presented quite an identification challenge – but a combination of forearm measurements and teeth examination eventually led me to land on *erythromus* for both of them. As dinner time approached, I could see from quite a distance I'd caught something else, something with white on its head. My brain frantically flicked through the pages and pages of bat images in my head, looking for what it could be. When I grabbed the bat to extract it, it became immediately apparent this was an *Anoura* that had a big patch of pollen on its forehead, evidence of a recent meal. Unlike the one I'd caught at Bellavista, this one had no tail at all. For this very straightforward reason, I decided it was *Anoura peruana*. Coincidentally, as I was processing the cutie with the pollen patch, the guide whom I'd met at Bellavista came to check on the bats I'd caught and I was able to show him the differences between this one and *Anoura cultrata*.

While I was waiting for bats to get caught, Daniel, a member of staff, came to tell me of a bat roost in a nearby building and said he'd take me there the next day. The walk was longer than I imagined it would be but we did some birdwatching along the way, looking at the many colourful tanager species typical of Andean cloud forests. The many tapir tracks along the way, belonging to a couple of Mountain Tapirs, a mum and her calf, gave me hope we perhaps could find one crossing a trail, but that didn't happen. No bears were seen either. What was seen, however, and what I actually came for, were bats. The abandoned building Daniel had told me about did indeed have bats hidden away underneath the remaining roof panels. After taking a few photos in situ, I took one out, examined it and measured it. It turned out to be a male Montane Myotis *Myotis oxyotus*, and I made sure to record it before letting go of my 368th species of the year. The species is relatively common in the Andes, but I'd been unable to record it thus far. I'd had suspicions from a few sequences recorded in Bellavista, but these bats were irrefutable proof of the species. This concluded a healthy selection of Andean specialities, both birds and bats. My only regret was that I didn't get to see the Torrent Duck despite this lodge being a reliable site for it.

Rio Claro, Costa Rica, December 2019

Fiona Reid, a famous mammal expert, illustrator and author, purchased a piece of land in southern Costa Rica, on which she built a lodge, Sylvan Waterfalls. While she'd had a few people visiting in the past, she wanted to organise a Bat Blitz for the grand opening. Like many others, she'd heard of the Big Bat Year, and she'd got in touch with me over the summer to make sure the dates coincided. I'd made plans to drive south with Ed and Sanne, a couple of bat workers who'd just moved to Australia from the Channel Islands. I would have had a good night's sleep, 'had it not been for the centipede that started crawling on my face in the middle of the night. I've woken up with the feeling of something crawling on my face a few times in life. It was never real, though. It was an entirely new experience for me to reach and actually grab something! The centipede was only about 4 cm long, far from the giant centipedes that are found in various places around the world (and that I'd seen in New Zealand in 2016) – but still, not a very fun experience. I went back to sleep shortly after this unfortunate incident, from which both of us, the centipede and I, escaped unscathed. At least physically; mentally, I don't know. I imagine the feeling from the centipede's point of view would be similar to me walking over the face of Godzilla when it's sleeping and then being grabbed by it.

The next morning, I joined Ed and Sanne again and we met up with Cecilia, Huma and Rhianna, all British bat workers. I'd chatted with Cecilia previously, as she was doing the same Mres I'd done the year before. We quickly took to the road as the drive ahead of us was long, but we were also on a fairly tight schedule. Fiona had told us to meet her and Jesus, the driver, in Rio Claro. She suggested we stop in Parque Nacional Carara for the Northern Ghost Bat *Diclidurus albus* (not to be confused with the Ghost False-vampire from Australia). We decided hiring a guide was the safest bet to increase our chances of finding this very special bat, especially given how little time we had, about one hour. Split between all of us, the cost was very reasonable, about US$15. That's probably the lowest amount 'I've ever paid for one special bat species. Our guide quickly understood that we were mostly interested in bats, so while he did give us some general wildlife information, he also skipped a lot of the generic stuff that he correctly assumed we already knew. He knew of a couple of areas that were good for *Diclidurus*. The first bats we saw were Greater Sac-winged Bats *Saccopteryx bilineata*, a common species that a few of us had already seen, but the rest of the group was delighted to get their first taste of Costa Rican bats. Less than a kilometre further down one of the trails, we entered the first area that was known to be reliable for the species we were after and before long, I spotted the first *Diclidurus*. We were all ecstatic as it is a tough species to find elsewhere, and it's not one we could hope for at Sylvan – it was now or never and even on a tight schedule, we'd found one. We all took some photos, but the bat was rather far away, hanging from the underside of a large palm frond (the technical term that applies to palm trees and ferns, as they don't have leaves).

After the obligatory photo with Kiwi, who was featured in most of the shots of rare bat species or breathtaking scenery, we continued down the trail, hoping that maybe we could find some more. And we did! We found two more under a different kind of palm frond, right above the trail this time. Unfortunately, at this point I 'forgot to pick up Kiwi and put it back in the bag'. That was the end of the road for my faithful companion. I hope a kid found it and still cherishes it to this day, rather than it ending up in a bin. As we didn't really have time to visit the rest of the park, we headed back to the cars, spotting an agouti along the way – a tick for everyone else but me as I'd already seen the Central American Agouti in Yucatán. We were even lucky enough to hear and see the Scarlet Macaws the park is known for, on top of finding three *Diclidurus* in one hour. I don't think anyone could have argued our short stop was not a resounding success. However, we couldn't stay there much longer as we had a train to catch. We did stop for a short lunch; well, it would have been short had it not been for the slow service. It gave us the opportunity for some birding on a mudflat. For a few of the other participants, this was their first time birding in the Neotropics, so they saw quite a lot of new species, including waders. Despite my limited experience with them, I was still able to assist in identifying a Willet, for example.

After finally reaching the lodge, we met with Christiane, Jonathan, Erik, Joris, Eva, Loren and Mike as they had already been catching bats for a little while. This day had been exhausting, but the various nets that had been set above the stream just outside the lodge were far too enticing for me to go straight to bed. Quite notably, I was the only one excited at the sight of the Proboscis Bat *Rhynconycteris naso*. They had caught a couple of them already, and most of the group had seen them elsewhere as well. However, I was still frustrated to have missed them in Peru and so I was relieved to be able to 'bag' that one, which became species #373. I did not have to worry any more about missing it entirely. I couldn't contemplate the frustration of missing the most readily seen

The last photo with my travel companion, Kiwi.

bat species in the Neotropics, one that every birder, every mammal-watcher, every nature enthusiast sees while travelling in Central or South America.

The other bats we were catching were *Molossus* species and various Phyllostomidae. Most of them I'd already seen, but I did want to make sure I didn't miss anything. It can be very chaotic when a dozen or so people check nets and process bats. Luckily, everyone knew what I was up to, and people regularly came up to me with a bat or two, asking me if I'd seen that species before. Releasing the *Molossus* we'd caught was quite funny as I knew from my time in Yucatán that they had to be thrown up in the air to help them take off, but most people present for the workshop weren't aware of that and started wondering what I was doing. It turns out that this is genuinely the only way to do it, and after Fiona had backed me up on that, we all started throwing *Molossus* into the air as we were capturing quite a lot of them. We were mostly catching *Molossus rufus*, confusingly called the Black Mastiff Bat. I know – I don't understand it either, and it's my favourite example to cite when people ask me why I don't love using common names. They can indeed be jet black, a gorgeous shiny jet black actually, probably my favourite colour on a bat, but the scientific name and the common name shouldn't contradict themselves; that's confusing.

One individual caught our attention as it was smaller and browner in colour; it turned out to be a Miller's Mastiff Bat *Molossus pretiosus*, which was confirmed after much debate and people checking and double-checking measurements and teeth structure. The issue with a lot of Neotropical molossids is that they're rarely the target species of trapping surveys, and it's rare to be catching them at normal net heights. Here, we were using a 'triple-high net' that does exactly what the name suggests – for bat people, confusion is only acceptable when it comes to species names, not equipment – it's three nets stacked on top of each other with a clever mechanism to bring them up and down to extract the bats. These nets are incredibly expensive and therefore unobtainable for most bat researchers, leading to a significant gap in what we know about the distribution of a number of Molossidae species, such as *M. pretiosus*. Making sure we weren't misidentifying it was key, but so was making sure we all understood the identification so that if in the future we were to catch one again, we'd know what it was. Rare species like that are incredibly interesting to catch because it really forces everyone to be methodical about the identification, rather than jumping to conclusions, and it's also a chance to learn something new. I tried to fit it with a light tag to collect reference calls, but this didn't work, so there are still no good reference calls for the species in Costa Rica. Most Molossidae roost high up in trees, making the collection of reference calls difficult, to the point that getting usable calls from hand-released bats is the only way. This method, however, is rarely suitable for fast and high fliers like this one, because they're unlikely to be behaving normally after release; by the time they have started behaving and calling normally, they're usually long gone. This is where the light tag could have helped track the bat for a little while longer, but I had no prior experience with the method. My attempts to get one to stick on a couple of *M. rufus* to try it out were unsuccessful, and the bat had to be released.

I suspect that for most people, the highlight wasn't *Rhynconycteris* but rather the famous Common Vampire Bat *Desmodus rotundus*. These can be highly unpredictable when it comes to catching them in nets. They are sporadic in undisturbed habitats, but not every farm with animals has them either. In this case, it's hard to know why the bats were there. Was it used to feeding on wild mammals or cattle? Whatever the case, this was a pleasant surprise and one that the group welcomed as none of them had caught or even seen one before. Needless to say that a few photos were taken of this animal! Luckily, vampire bats are very sturdy and tend to be easy to handle even for relatively extended periods of time. Even then, the bat was released as soon as possible, and it showed us their typical way of flying away upon release – that is, dropping to the ground and running off.

The following nights were all packed with bats as we tried different habitats, sampling other parts of the land Fiona owned. On average, we were getting two or three new species for my list every night, which was incredible. From the cute tiny Northern Little Yellow-eared Bat *Vampyressa thyone*, to the not so tiny Heller's Broad-nosed Bat *Platyrrhinus helleri*. We also caught the rare Velvety Fruit-eating Bat *Enchistenes hartii*, a highly distinctive Phyllostomidae – there aren't too many of those that are of a beautiful rich brown colour with buffy, almost golden in fact, stripes on the face. A truly gorgeous bat. Our trapping sessions became increasingly complicated with sometimes up to three processing stations for the common species, but the rare ones all made it back to the main station where I was staying, for obvious reasons, and Fiona was going back and forth to check on identifications. This complex scheme was rendered even more problematic when it started raining, and we all had to keep our equipment and the bats nice and dry while making sure everyone saw every species. This was a genuinely challenging task. Luckily, we all had the bats' best interests in mind, and everybody was happy.

My fatigue levels were getting dangerously high, though, and I became a bit cranky at times. Another issue that tends to arise when I'm tired is that my knee starts hurting again, as it has been on and off for the past eight years. One night when it hurt so much that I could barely walk, I decided not to torture myself that evening and to stay at camp while everyone headed up the ridge to catch bats up there. The climb was rough, which comforted me in thinking I'd made the right decision. Something completely unexpected ensued: a couple of people brought down a few bats for me that evening, so that I could see and examine them and keep adding new species to my list! The group's empathy was incredible. In the end, one of the species that would have been new for me wasn't brought down because of a miscommunication, and Fiona felt bad about that. Truthfully, I really couldn't have cared less, as I was already so touched by the fact that four bats had been brought down for me.

Peter's Disk-winged Bat *Thyroptera discifera* was a particularly sought-after species for all of us. Fiona had found a roost shortly after she bought the property and she'd managed to keep track of their whereabouts afterwards. When Jon Hall, another famous mammal-watcher and definitely an inspiration for many of us when it comes to mammal-watching travels, came to visit, they'd successfully located them again. They had been left alone until we showed up. It took the whole team, deployed across the entire banana plantation, to find them once more. They were, as usual, all huddled up in what appeared to be one continuous ball of fuzzy bat butts. Catching them for measuring and weighing turned out to be a much more significant challenge than anticipated. We all expected they'd fly out as soon as we touched the dead leaf they were roosting inside, but they took quite a lot of convincing to make their way into the hand net. We hadn't disturbed them for nothing though. Precious little is known about their roosting habits, particularly regarding age and sex, and this was the perfect opportunity to add to that scanty data. The roost comprised of seven individuals, six males and one single female. They averaged a weight around 3 g, which isn't all that much more than the weight of the average Bumblebee Bat. They were spectacularly tiny. Thanks to the fact that there was a few of them to process, everyone got a chance at handling one and studying it in detail. This way, we avoided over-stressing any one individual by having to pass them along.

For me, the highlight of the trip came on one of the last trapping nights, one where I was being a bit lazy and wasn't contributing to the trapping effort much; but then things got hectic, and we needed people extracting constantly, and then other people would ferry the bats from the nets to the lodge for processing. I don't think I'd ever had such a busy night of trapping. All the bats waiting to be extracted had been shouting, as Phyllostomidae always do, and this can attract

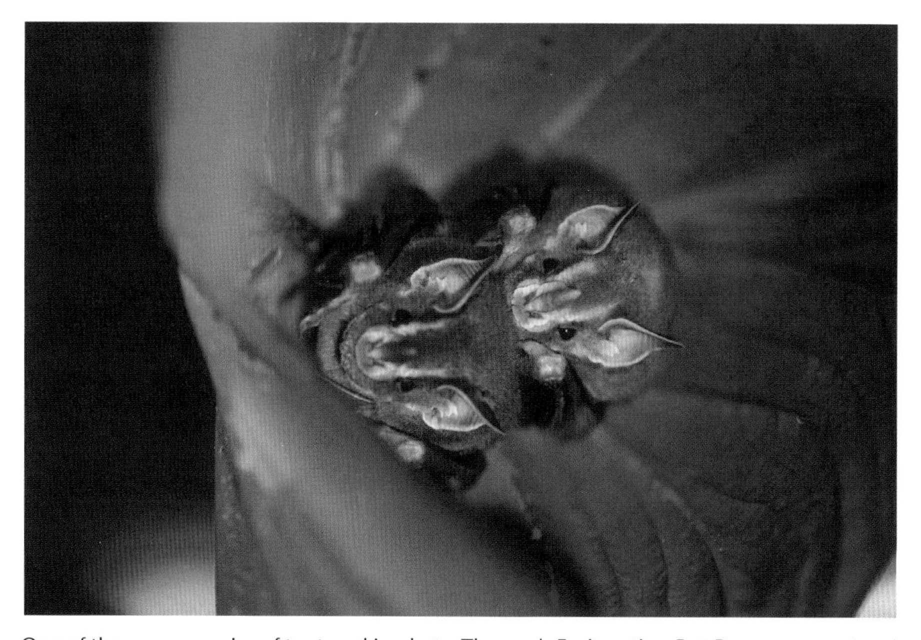

One of the many species of tent-making bats, Thomas's Fruit-eating Bat *Dermanura watsoni*.

other bats. It's not known exactly why this happens. Is it because other bats come to see if they're in a position to help? Altruism has been described in bats, most notably vampire bats. Or is it the equivalent to humans when there's an accident on the motorway that causes a traffic jam in both directions, because people on the other lane can't drive straight and feel the need to slow down to see what happened? Other bats aren't the only thing attracted to the sound of bats squeaking in the nets; predators are too. In this case, an Ocelot had come to see if there was a real opportunity. It would not be given the chance at a meal, however, because Fiona was making sure the bats were safe, and she also ensured that we could see the cat – calling us just before it came back for another look or two just to make sure it wasn't missing out on anything juicy. Those nets were closed shortly after as the capture rate was untenable, and the presence of a roaming cat was suboptimal for the bats' welfare. In just a little over a week, we'd found over 50 bat species, of which 18 were new for my list, putting me within spitting distance of the magical 400 mark.

The days were relatively quiet and, for me, were mostly spent in looking at the recordings from the night before. Still, we also took a day trip to Corcovado National Park, where we did see a few new bat species (of course!), such as Thomas's Shaggy Bat *Centronycteris centralis* perched on a tree – cracking views of it actually. Checking every termite mound we could find happened to yield a group of Pygmy Round-eared Bat *Lophostoma brasiliense*, which I hadn't seen yet. In Peru, I'd got acquainted with a closely related species, White-throated Round-eared Bat *Lophostoma sylvicolum*, but it was a distinct species. I was getting seriously confused by all the similar names though. Other notable mammals included a brief sighting of the Northern Tamandua. After missing its southern cousin, I was delighted to still have an opportunity to see one of these anteaters. Unfortunately, in true Big Bat Year fashion, we did see tracks of tapirs (that was all of them ticked off as far as tracks were concerned), but there was no actual tapir to be found.

Manu joined me for the last week of the year, the final stretch when we visited two more reserves, Tirimbina and Monteverde. The main target at the former was the world-famous Honduran White Bat *Ectophylla alba*, only found on the Caribbean side of Costa Rica. On the way to the first reserve, we made a brief stop to add the American Dipper to our bird list. Tirimbina is a well-known place, famous for its sloths, birds and also because this reserve makes it possible for many travellers to see rare species. One thing they focus on, which is rather unusual, is bats. The most popular species there is undoubtedly *E. alba*, but the reserve is one of the best locations in the country for a number of emballonurid species as well. Once we were checked in, there was no time to lose and we immediately headed out on a walk with Carolina, a scientist working at Tirimbina, to look for *Ectophylla*, on Christmas Day. It really helped that she already knew exactly where the bats were – and before I'd realised what was happening, we'd found the leaf, surprisingly low to the ground, with the cute white balls of fluff inside. The leaf was too close to the ground for me to take photos with my proper camera, which coincidentally had stopped working anyway, likely because of the constant downpour we'd been under since we arrived in at Tirimbina. Manu's phone almost died as well. There was no doubt as to why it's called the rainforest! Before Carolina had to go back to work, she was able to show us a roost of the Chestnut Sac-winged Bat *Cormura brevirostris*. These shy emballonurids roost on the underside of fallen tree trunks. We had one more bat to find in the reserve, the Short-eared Bat *Cyttarops alecto*, also an emballonurid – but that one, we had to plan a short trip for. This species has very different roosting habits to *Diclidurus* and *Cormura* as it typically roosts under palm leaves, high up above ground.

While we were there, Manu and I wanted to join some evening walks so that we could access the reserve at night, looking for bats, other mammals and whatever critter decided to show up. The rainforests of Costa Rica are vibrant, especially when it comes to amphibians. Being from Western Europe, I'm used to a ridiculously low level of amphibian diversity. As a result, going out in the evening to look for them wasn't really something I'd done much. But in Peru, I discovered how fun it could be, how little experience I had and how many critters I missed while walking around. While in theory, one wouldn't see anything different while on a guided walk, the experience of the guide is a precious asset when it comes to finding things like frogs and snakes. We showed up at the reception at 7 pm, slightly surprised to see so many other people there too. After an entire year travelling to some of the remotest places, we'd forgotten what it meant to be in a touristic area. At the same time, it was also good to see people interested in wildlife to the point that they wanted to go on a night walk.

A little further down the track, we stopped for a Hognosed Pit-viper *Porthidium nasutum* that was having a rest on the footpath. They tend to stay in the same spot for several days, and when they decide the trail is the most convenient spot, the rangers have to put signs around them because their bite is rather unpleasant. They're not particularly aggressive, but no snake, or no animal at all for that matter, enjoys being stepped on. A camouflaged snake would be easily trodden on if sticks didn't surround it to make its presence obvious. Not five metres further on, our guide showed us a Brazilian Wandering Spider *Phoneutria boliviensis*. It used to hold a Guinness World Record for being the most venomous spider in the world but was recently dethroned by the Sidney Funnel-web Spider. Luckily, this one isn't aggressive either, but it was best to keep our distance regardless. Two metres beyond the deadly spider, a couple of Bullet Ants *Paraponera clavata* were on a stroll. At that point, Manu and I looked at each other and knew this trail should be called 'the Trail of Death', for obvious reasons. While Bullet Ants aren't deadly, they can cause a lot of pain, even if their reputation of having a sting as painful as a gunshot is highly exaggerated, as are many things in the jungle. Luckily, I escaped all of their attempts to show me what they're capable of. It looked like a rather interesting succession of somewhat dangerous animals – a bit like how most people tend to picture Australia.

In addition to the night walks, which turned out to be incredibly productive, we also attempted some mist-netting, which in contrast was anything but productive. We were catching bats but almost exclusively *Artibeus jamaicensis* and not the Wrinkle-faced Bat *Centurio senex* we were hoping for. The few fruiting fig trees that we'd set up our nets near to looked promising, but there didn't seem to be any Wrinkle-faced Bats out that night. Our trapping session was ended abruptly following the attacks of a Grey Four-eyed Opossum on our nets. It had already killed one bat and was about to go for another by the time we got to the net and were able to chase it away. As we got back to the second net, after closing the first one to avoid further death or even stress to the bats, the opossum had found it too and got dangerously close to these bats as well, so we closed it too. We were told that *Centurio* can be a late-night species but the risk of predation by the opossum was too great for us to continue trapping. I thought it would have been an unethical thing to do. Interestingly, the Grey Four-eyed Opossum is a species that the team of Dutch mammal-watchers we'd met at Sylvan had missed, despite their many attempts. What I wouldn't have given to not have to deal with it, and to be able to mist-net in peace...

The final species of the BBY: Toltec Fruit-eating Bat *Dermanura tolteca*.

When we got to Monteverde, Vino, a fellow Belgian bat researcher now living in Costa Rica and managing the Bat Museum there, found himself meeting a very tired Nils who had to be strongly encouraged to do anything. My energy levels were at an all-time low and although this was the last stretch and I knew I had it give it everything I had, that wasn't much. We'd planned two trapping sessions, one inside the reserve and one outside. On the evening outside the reserve, we focused on an area with picnic tables and hummingbird feeders. Our target was Geoffroy's Tailless Bat *Anoura geoffroyi* – because of the feeders, not so much because of the picnic tables. Had I done my Big Bat Year a few years earlier, all the *Anoura* I would have seen would have been considered as being the same species. Many *Anoura* species have recently been split from *geoffroyi*, including all the species I'd seen so far; as a result, the 'parent' species was still missing from my list! We caught a couple of them, right as people on a night walk passed by our nets, enquiring about our ongoing activity. We took the time to explain the workings of bat research, what kind of information we could get from trapping, and people got a fantastic bonus to their walk in the form of a close sighting of a nectar-feeding bat.

After a trio of *Sturnira* species we caught inside the reserve during a rainy trapping session, I was running out of new species I could reasonably target in the area, with one notable exception: the Toltec Fruit-eating Bat *Dermanura tolteca*. Luckily, we didn't have to wait to catch one; Vino knew the owner of a house where a few were roosting under the roof of the terrace. The owner was more than happy to show us what ended up being the last species of the Big Bat Year – the 396th at the time, but the 400th as I type thanks to taxonomic changes and one identification revision based on photos.

Bats and ecotourism

Birds are a well-known driver for tourism; an estimated three million international trips take place per year with their main focus on birding. It's estimated that over US$40 billion are spent on birding every year in the United States of America alone. Clearly, the incentive is there for birding lodges, private reserves and so on. A financial incentive makes it easier for people to get involved in conservation.

What about bats? Can ecotourism help them too?

While bat-driven tourism is very unlikely to reach the same value or the same level of enthusiasm as the birding industry, it is worth something nonetheless, and its value needs to be acknowledged. Even though bat-watching may only be the main purpose of travel for roughly three and a half people a year, the sites offering bat-watching opportunities play an important role in bat conservation, even if they are not huge sources of revenue, especially compared to birding sites.

Bracken Cave is home to the largest bat colony. Unlike many other cave roosts worldwide, the only species present is the Mexican Free-tailed Bat *Tadarida brasiliensis*. The area around the cave has been modified to accommodate benches for the comfortable viewing of the emergence of millions of bats (up to 15 million according to some estimates), and fences were installed to prevent visitors from entering the cave. By becoming members of Bat Conservation International visitors help fund their research and conservation projects around the world, and in exchange, they get to see one of the greatest wildlife shows on Earth. If you're after a group experience, tours run by Merlin Tuttle Bat Conservation allow anyone to see some of the most amazing bat species in the world, including in Thailand and Ecuador. Those tours are incredible ways to support local communities who protect bats, as well as Merlin's organisation.

At the opposite end of the scale of bat ecotourism is a plethora of Asian villages where locals take tourists to visit a flying fox colony or a free-tailed bat colony for a minimal fee that provides them with additional income. While the funds don't directly go towards bat conservation projects, they ensure the survival of the roosts that locals show to visitors. Many Westerners on 'adventure' holidays seek close interactions with 'exotic' animal species, and bats fall into that category. They are unlikely to have encountered giant fruit bats or bats in high quantities. This is especially true given that most tourists are probably coming from regions where bats have a bad reputation. Places such as Gomantong Caves in Sabah, the Painted Bat village in Thailand or the Sulawesi Flying Fox colony in Sampakang can help change that. Changing people's perceptions of bats is arguably as important as any other part of conservation, even through international travel, with its inherent social and geographical inequities.

Finally, ecotourism does not necessarily have to revolve around focus species; charismatic species can be used in marketing materials for night walks, for example, but these are fantastic education opportunities to provide participants with information on lesser-known or misunderstood species such as bats, and their importance in the ecosystem. Ecotourism isn't only about raising funds; it's also an invaluable outreach opportunity that's too seldom taken advantage of.

Brussels, Belgium, January 2020

This rollercoaster of an adventure had come to an end. I could hardly believe it. I struggled to return to normal life; I had no real plans for what I'd do in the following months. Soon after, a pandemic hit the world, leading many people to tell me how lucky I was to have been able to do the Big Bat Year right before the world went into lockdown. I wasn't just lucky to have escaped a lockdown, I was lucky to have been able to go on such a journey, see so many bats, meet so many people, enjoy so many different foods and learn about so many different cultures. I experienced so much in so little time. It's hard to believe how quickly everything went by. From deserts to mountains, not mentioning the countless types of forests I visited. Equally impressive was the diversity of bats I encountered, from the two-gram Bumblebee up to the one-and-a-half-kilogram giant flying foxes in Southeast Asia. Some were fruit-eating bats, some carnivorous; others were fishing bats and some others even fed on blood. Bats aren't all small, brown and dull. They are a hugely diverse group, and I was able to get an incredible overview of them. And those are just the highlights of the bat species I saw, without even mentioning the 2,500+ bird species, countless reptiles, amphibians and other mammals – and of course, all the landscapes that will forever be engraved in my brain.

It's hard to talk about such a journey without mentioning the elephant in the room: the carbon footprint. With the carbon footprint of that of a small town, I can't deny I haven't contributed, in some way, to climate change. But we all do. Without wishing to make it sound like I don't take this issue seriously, it's important to acknowledge that while sky travel is the most significant source of emissions by private individuals, at a worldwide scale, it is dwarfed by industries and indeed, the internet. Therefore, I think this journey's impact on the planet must be evaluated by weighing the positive and the negative. But what are those positive impacts exactly?

Can discovering a new species in a remote part of Indonesia be considered a valuable contribution to bat conservation? Probably not! However, the impact of the Big Bat Year goes well beyond the few scientifically meaningful discoveries I made (which at the time of writing, I haven't even written up yet for publishing...). My journey also appealed to birders; some reached out to me about the best way to get into bats. A couple of bat workers reached out to me to pick my brains on starting their own Big Bat Years, though restricted to a single continent. I'm still getting regular requests from people to assist them in their travel plans.

One could say that the tables turned somewhat. I went from wanting to see as many bat species myself to wanting to show as many of them as possible to as many people as possible. Because of Covid restrictions, a lot had to happen online – but that's when I realised how valuable all the photos I'd taken were. Luckily, in between lockdowns, I was still able to share stories and my knowledge of bats in person with people of all levels of experience, which is what I love most about outreach: it's the diversity of the audience. From my very first talk on the Big Bat Year at the British National Bat Conference in September 2019, I've loved every single chance I've had to share my stories and to discuss bat conservation with people from all backgrounds.

Why is travelling important for bat conservation? Travelling inspires people. It inspires people to follow in your footsteps, yes, but it doesn't stop there. I've met countless people who, hearing about stories of bats around the world, wanted to do more for the bats in their garden. Many have told me that they were inspired by my journey, that it made them dream. I'm a firm believer that while science is important in conservation, so is dreaming and optimism. That part is often overlooked, and it is my belief that my journey was a healthy mix of both science and dream. As a result, the Big Bat Year became the seed for a project focusing on outreach, albeit at a much larger scale: International Bat Night. Thanks to the many connections I was able to make during the BBY, I was able, in 2021, to organise the first ever worldwide live event celebrating bats and the people who work with them. Bat rescuers, researchers, photographers and enthusiasts all shared their life with bats with the public. The pandemic has normalised online events and they're a great way to make science accessible to everyone with an internet connection. This is the goal of International Bat Night, combining science and passion for bats to bridge the gap between scientists and the public.

Why did I decide to write a book? I wanted to be able to share my experience of travelling around the world to see bats with as many as possible. I had so many stories of bat conservation, both good and bad, to tell. While my talks allowed me to share a great many of them, I felt this wasn't enough. I also kept getting requests from people to write a book, so I eventually decided to get in touch with Pelagic Publishing and here we are!

Would I do this again? Probably not. I will continue to travel to photograph and research bats, but not at the same pace as the Big Bat Year, which brought frustration and exhaustion. But this kind of challenge is also extremely rewarding and fun, and I would love for someone to follow in my footsteps. Why not beat my record? I 'missed' over 1,050 species (at the time of writing, this number goes up every month as new species are discovered) – that leaves a lot of room for someone to build on what I did. Reach out to me if you are interested in doing this; I'd love to help you by sharing what I think I did well and what I think I did wrong. My personal goal will be to reach 1,000 bat species by the end of my lifetime.

As I'm writing these lines, over two years after the end of my journey, I'm often asking myself the question of where I'd be without this world tour. Where I'd be if I hadn't gone to the UK for my Master's. It's easy to ask ourselves where we would be if we hadn't done this or that, but this world tour, and the Master's to a lesser extent, really seems like a life-changing event. I guess the best evidence of that is the fact you wouldn't be reading this if I hadn't one day woken up in my bunk bed in the Canary Islands and decided to make the Big Bat Year really happen.

Acknowledgements

Beyond the life-changing aspect of the journey, there's something else I like to reflect on – the jaw-dropping number of people who supported me on this journey. From people who gave me information on where to find a specific bat cave to people who went out of their way to make my journey easier, or who provided invaluable emotional support during what was a very challenging 12-month mental health journey, I owe them all thanks. That is why I would like to dedicate the next pages to those Big Bat Year heroes:

New Zealand: Abi Quinnell and other DOC volunteers, Mike Ashbee
Fiji: Dave Waldien
Solomon Islands: Alistair and Richard and my guides on the waterfall journey. They saved my life and my equipment.
New Caledonia: Isabelle Jollit and Lucie
Australia: Damian Milne, Alan Gillanders, John Harris, David Nixon, Jenny Maclean, Vanessa Gorecki, Luke Hogan, Stuart Parsons and Oscar Bassem
Philippines: Will Cabanillas and Bram Demeulemeester
Indonesia: Carlos Bocos, Vinno Swoerlan, Irawan Halir, Arjan Boonman and the owner of the homestay on Waigeo
Malaysia: Nur Izzati Abdullah
Thailand: Daniel Heargreaves, P'Kwang and his nephew
Nepal: Sanjeev Baniya, Emily Stanford Miller and the couple who kindly let us into their home when we found ourselves caught in the rain with no shelter and no way to go home
India: Rohit Chakravarty, Harpeet Kaur, Rajesh Puttaswamaiah and Rajkumar Patel
Taiwan: Joe Chun-Chia Huang, Richard Foster and Keith Barnes
Japan: Jason Preble
USA: Merlin Tuttle, Teresa Nichta, Lee Mackenzie and Dianne Odegard
Mexico: Lorenzo Falcao, Luke Eberhart-Phillips, Juan Cruzado Cortes, Jesper-Bay Jacobsen
Madagascar: Daniel and his son, and my taxi driver

Kenya: Paul Webala, Kevin Lemantaan, Sylvie, Ben, Jean-Baptiste, Clément and Pauline
Israel: Arjan Boonman, Noam Weiss and Yoav Perlman
Europe: Joe Szewczak and Daan Dekeukeleire
Peru: Chris Ketola, Chris Kirkby and Laurie Allnatt
Ecuador: Richard Parsons, Antonio, Martin Conyers and Amanda Gee, Irene de Cunca Costa, Gabriel and Daniel
Costa Rica: Loren Ammerman and the rest of the blitz team (Cecilia, Jonathan, Christiane, Ed, Sanne, Huma, Rhianna, Mike) and the Dutch folks who briefly spent time with us at Sylvan, Vino De Backer and David Rodríguez Arias

I would also like to extend my most sincere gratitude to Fiona Reid and Jon Hall for their help throughout my journey, as well as Ellie Huckle and Emily Stanford Miller for their support, despite the circumstances. Manu, who joined me in New Zealand, Thailand, Kenya and Costa Rica, I am also very grateful for your support and your companionship.

Merlin Tuttle has been inspiring me greatly since the beginning of my journey with bats and without him, I'm not sure I would ever have understood the significance of photography for bat outreach, or the importance of travel. And without Arjan Dwarshuis, there wouldn't have been the original idea to embark on such a voyage, nor the motivation to write a book about it. Thank you so very much to both of you for your invaluable contribution to my Big Bat Year.

Thank you to all those who supported my crowdfunding campaign on GoFundMe: Laure, Ellie, Arjan, JoEllen, Olivier, François, Renata, Marianne, Martine, Corinne, Florence, Franck, Viviane, Arno, Serge and Laurel.

Finally, this journey couldn't have happened without my parents, Philippe Bouillard and Arielle Delhaye, as well as my sister Selma Bouillard and my grandmother Denise Henderickx. Their unwavering support of my craziest endeavours is what has allowed me to live my passion to the fullest.

Additionally, I'd like to thank those who provided feedback on the manuscript: Stuart Newson, Merlin Tuttle, Jonathan Townsend, Rob Thomas, Ellie Huckle, Steven Allain, Rhiân Ebrey and Soetkin De Vos, as well as Erin Chamberlain and the Write Now! members.

Index

Acerodon celebensis (Sulawesi Flying Fox), Tangkoko, Indonesia 60, *61*
Acerodon leucotis (Palawan Flying Fox) 51
Andasibe, Madagascar 142–45; Diademed Sifaka 143–44; *Miniopterus* and *Scotophilus* species 142; *Miniopterus majori* (Major's Long-fingered Bat) 143; Mouse-lemur 145; *Neoromicia matroka* (Malagasy Serotine) 143; Ring-tailed Lemur 142; *Scotophilus robustus* (Robust House Bat) 145
Anoura cultrata (Handley's Tailless Bat), Galápagos, Ecuador 186
Anoura geoffroyi (Tailless Bat), Galápagos, Ecuador 201
Anoura peruana (Peruvian Tailless Bat), Galápagos, Ecuador 186, 190
Anoura species, Galápagos, Ecuador 201
Antrozous pallidus (Pallid Bat) 179
Antsiranana, Madagascar 139–41; baobabs 140; *Eidolon dupreanum* (Madagascan Straw-coloured Fruit Bat) 140; *Eidolon helvum* (African Straw-coloured Fruit Bat) 140; *Mops leucostigma* (Malagasy White-bellied Free-tailed Bat) 139; *Myotis goudotii* (Malagasy Myotis) 141; Northern Rufous Mouse-lemur 140; *Paraemballonura tiavato* (Western Sheath-tailed Bat) 140; *Tadarida fulminans* (Malagasy Free-tailed Bat) 139; *Triaenops menamena* (Rufous Trident Bat) 140–41
Aquatic Warbler, Virelles, Belgium 114
Armenian Gull, Eilat, Israel 166
Armstrong, Kyle 24
Artibeus jamaicensis (Jamaican Fruit-eating Bat), Galápagos, Ecuador 200; Yucatán, Mexico 129
Asellia tridens (Trident Leaf-nosed Bat), Tel Aviv, Israel 163
Aselliscus stoliczkanus (Trident Bat), Kaeng Krachan, Thailand 85
Aselliscus tricuspidatus (Temminck's Trident): Tangkoko, Indonesia 62; West Papua, Indonesia 64
Attenborough, David 120, 169
Austin, USA 116–20; Bracken Cave 117–19; caring for bat pups 117; *Dasypterus xanthinus* (Western Yellow Bat) 119; *Eidolon helvum* (Straw-coloured Fruit Bat) 118; *Kerivoula picta* (Painted Bat) 118; *Lasiurus borealis* (Eastern Red Bat) 118; Merlin Tuttle's Bat Conservation 116; *Myotis lucifugus* (Little Brown Bat) 118; *Myotis velifer* (Cave Myotis) 118; *Nycticeius humeralis* 119; 'Pipistrelles' 119; Ruby-throated Hummingbird 119; *Tadarida brasiliensis* (Mexican Free-tailed Bat) 118, 119
Austronomus australis (White-striped Free-tailed Bat), Queensland, Australia 36
autism spectrum disorder (ASD) 91

Balantiopteryx plicata (Grey Sac-winged Bat), Jalisco, Mexico 127
baobabs, Antsiranana, Madagascar 140
Barbary Ground Squirrel 2
Barbastella barbastellus (Barbastelle Bat), Chengdu, China 95
Barbastella darjelingensis (Darjeeling Barbastelle), 106
Barbastella leucomelas (Eastern Barbastelle), Eilat, Israel 165
Bassaricyon neblina (Olinguito), Galápagos, Ecuador 187

Bat Blitz, Rio Claro, Costa Rica 192
bat conservation community 69
Bat Conservation International 116
Batman 111
Bengaluru, India 88–90; *Hipposideros galeritus* (Cantor's Leaf-nosed Bat) 88; *Lyroderma lyra* (Greater Asian False-vampire) 89; *Otomops wroughtoni* (Mastiff Bat) 88; *Rhinolophus beddomei* (Horseshoe Bat) 88–90; *Tadarida aegyptiaca* (Egyptian Free-tailed Bat) 89
Biebrza National Park, Virelles, Belgium 115
Big South Cape Island, New Zealand 29
Birdfair (Britain) 5
Black Grouse, Galápagos, Ecuador 187
Black-necklaced Scimitar-babbler, Taiwan 107
Blue Boobies, Galápagos, Ecuador 182
Blue Duck/Whio, New Zealand 8
Bocos, Carlos 63, 65, 67
Boonman, Arjan 64, 163–65
Bracken Cave, Austin, USA 117–19, 202
British National Bat Conference 205
Broad-billed Sandpiper, Eilat, Israel 166
Brussels, Belgium 204–6
Buff-collared Nightjar, Jalisco, Mexico 126

Cane Toads 14
Cape Ternay area, Mahé, Seychelles 161
Cardioderma cor (Heart-nosed Bat), Nairobi, Kenya 147–48, 179
Carollia perspicillata (Short-tailed Bat): Puerto Maldonado, Peru 174, 177; Yucatán, Mexico 130
Carollia sowelli (Sowell's Short-tailed Bat), Yucatán, Mexico 130
Centronycteris centralis (Shaggy Bat), Galápagos, Ecuador 198
Centurio senex (Wrinkle-faced Bat), Galápagos, Ecuador 187, 200
Chaerephon jobensis (Greater Northern Free-tailed Bat), Queensland, Australia 36
Chaerephon johorensis (Lesser Northern Free-tailed Bat), Taman Negara National Park, Malaysia 79
Chaerephon plicatus (Wrinkle-lipped Free-tailed Bat), Subic Bay, Luzon, Philippines 49, 53
Chaerephon plicatus, Taman Negara National Park, Malaysia 78
Chakravarty, Rohit 88
Chalinolobus morio (Chocolate Wattled Bat), Darwin, Australia 45
Chalinolobus tuberculatus (Long-tailed Bat), New Zealand 8, 9, 11, 12
Chengdu, China 95–97; *Barbastella barbastellus* (Barbastelle Bat) 95; *Myotis blythii* (Lesser Mouse-eared Bat) 96; *Myotis chinensis* (Large Myotis) 96; *Myotis dasycneme* (Pond Bat) 95; *Myotis myotis* (Greater Mouse-eared Bat) 96; Qinglonghu Wetland 95
Chironius fuscus (Brown Sipo), Puerto Maldonado, Peru 172
Christian beliefs 110
Christmas Island 29
Chrotopterus auritus (Woolly False-vampire Bat), Yucatán, Mexico 134
Cloeotis percivali (Short-eared Trident Bat), St Lucia, South Africa 156
cloud forest, Galápagos, Ecuador 188
Cock-of-the-Rock (COTR), Galápagos, Ecuador 187, 189
Coleura afra (African Sheath-tailed Bat), Nairobi, Kenya 149
Coleura seychellensis (Seychelles Sheath-tailed Bat), Seychelles 29, 160
Collared Flycatchers, Virelles, Belgium 114
Collared Owlet, Taiwan 106
Colo-I-Suva Rainforest Eco Resort 17
Cormura brevirostris (Chestnut Sac-winged Bat), Galápagos, Ecuador 199
Cortes, Juan Cruzado 129
Crab Plover, Mahé, Seychelles 160
Craseonycteris thonglongyai (Bumblebee Bat), Kaeng Krachan, Thailand 84
culture and bats 110–11; Christian beliefs 110; Dracula 111; Egyptian mythology 110; face of a bat, 'God of Death,' Camazotz 110; Mayas, Month of the Bat 110
Cynopterus brachyotis (Forest Short-nosed Fruit Bat), Taman Negara National Park, Malaysia 79
Cynopterus luzoniensis (Peter's Fruit Bat), Subic Bay, Luzon, Philippines *53*
Cyrtodactylus consobrinus (Banded Forest Gecko), Sepilok, Borneo 73
Cyttarops alecto (Short-eared Bat), Galápagos, Ecuador 199

Darwin, Australia 43–45; *Acanthophis rugosus* (Rough-scaled Death Adder) 43; *Chalinolobus morio* (Chocolate Wattled Bat) 45; ghost bat *44*; Golden Bowerbird 45; landscape *44*; *Macroderma gigas* (Ghost False-vampire Bat) 44; *Nyctophilus* and

Taphozous 44; *Rhinonicteris aurantia* (Orange Leaf-nosed Bat) 44; Short-tailed Brushtail Possum 45
Darwin, Charles 169
Darwin finches, Galápagos, Ecuador 181
Daxueshan National Forest, Taiwan 105
Demeulesteer, Bram 53
Department of Conservation (DOC) 10
Dermanura tolteca (Toltec Fruit-eating Bat), Galápagos, Ecuador 200
Dermanura watsoni (Thomas's Fruit-eating Bat), Rio Claro, Costa Rica 197
Desmodus rotundus (Common Vampire Bat) 86, 137; Jalisco, Mexico 128; Rio Claro, Costa Rica 195; Yucatán, Mexico 131
Diademed Sifakas, Andasibe, Madagascar 143–44
Diclidurus albus (Northern Ghost Bat), Rio Claro, Costa Rica 192, 193
Diphylla ecaudata (Hairy-legged Vampire Bat), Yucatán, Mexico 132
disease and bats 98–99; antibodies 98; bat virology 98–99; Covid-19, effect on 98; *Desmodus rotundus* (Common Vampire Bat) 99; MERS 98; *Rousettus aegyptiacus* (Egyptian Fruit Bat) 99; SARS 98; SARS-CoV-2 98; social distancing 99
diving break, Sulawesi, Indonesia 54–56; *Kyonemichthys rumengani* (Lembeh Pygmy Seadragon) 54; *Metasepia pfefferi* (Flamboyant Cuttlefish) 54, 56; Orangutan Crab 56; sea anemone *55*; *Synchiropus splendidus* (Mandarin Gobies) 54
Dobsonia inermis (Naked-backed Fruit Bat), Guadalcanal, Solomon Islands 26–27; West Papua, Indonesia 65
Dobsonia viridis (Greenish Naked-backed Fruit Bat), Tangkoko, Indonesia 61
Doppler effect 46
Dracula 111

echolocating bats, Queensland, Australia 34
echolocation 46–47; Doppler effect 46; frequency 46; *Macrophyllum macrophyllum* (Long-legged Bat) 47; Pteropodidae 47; quiet bats 47; *Steatornis caripensis* (Oilbird) 47; steep echolocation calls 46
EchoMeter 11, 37
ecotourism and bats 202–3; bat-driven tourism 202; Bracken Cave 202; flying fox colony or Free-tailed Bat colony 202; Gomantong Caves in Sabah 203; Merlin Tuttle Bat Conservation 202; Painted Bat village, Thailand 203; Sulawesi Flying Fox colony, Sampakang 203; *Tadarida brasiliensis* (Mexican Free-tailed Bat) 202
Ectophylla alba (Honduran White Bat), Galápagos, Ecuador 198
Egyptian mythology 110
Eidolon dupreanum (Madagascan Straw-coloured Fruit Bat), Antsiranana, Madagascar 140
Eidolon helvum (Straw-coloured Fruit Bat) 169–70; Antsiranana, Madagascar 140; Austin, USA 118
Eilat, Israel 165–68; Armenian Gull 166; *Barbastella leucomelas* (Eastern Barbastelle) 165; Broad-billed Sandpiper 166; conference on Big Bat Year 168; Eilat International Birding and Research Centre 166; *Eptesicus bottae* (Botta's Serotine) 165; *Eptesicus nilssonii* (Northern Bat) 167; K20 saltpans 166; mouse-tailed bats 167; *Nyctalus lasiopterus* (Giant Noctule) 167; *Pipistrellus maderensis* (Madeiran Pipistrelle) 167; *Plecotus christii* (Christie's Long-eared Bat) 165, 167; *Plecotus teneriffae* (Tenerife Long-eared Bat) 167; *Rhinopoma cystops* (Big-eyed Mouse-tailed Bat) 165, 167; spring migration 165; *Taphozous nudiventris* (Naked-rumped Tomb Bat) 167; *Vespertilio murinus* (Particoloured Bat) 167; wheatears 167; White-cheeked Tern 165; White-eyed Gull 165
Elephanta Caves, Mumbai, India 92
Elfin forests, Galápagos, Ecuador 189
Elliott's Storm-petrel, Galápagos, Ecuador 184
Elvish settlements 7
Emballonura semicaudata (Pacific Sheath-tailed Bat), Viti Levu, Fiji 21, 22
Enchistenes hartii (Velvety Fruit-eating Bat), Rio Claro, Costa Rica 196
Eptesicus andinus (Little Black Serotine), Galápagos, Ecuador 186
Eptesicus bottae (Botta's Serotine), Eilat, Israel 165
Eptesicus furinalis (Argentinian Serotine), Yucatán, Mexico 131
Eptesicus hottentotus (Hottentot Serotine Bat), St Lucia, South Africa 152

Eptesicus nilssonii (Northern Bat), Eilat, Israel 167
Eptesicus pachyomus (Mouse-like Serotine), Taiwan 104
Eumops ferox (Wild Bonneted Bat), Yucatán, Mexico 130
Eumops underwoodi (Underwood's Bonneted Bat), Yucatán, Mexico 133
evolution 75–76; fossils 75; *Icaronycteris index* 75–76; *Onychonycteris* 76; *Tachypteron* 76; *Tachypteron franzeni* 76

Fiji Bush-warbler, Viti Levu, Fiji 20
Flamecrest, Taiwan 106
Fogg Dam 44
food habits 178–79; *Antrozous pallidus* (Pallid Bat) 179; blood 178; *Cardioderma cor* (Heart-nosed Bat) 179; diverse diets 178; fishing 179; frogs 178–79; fruit-based 178; insects as primary 178; *Myotis* genus 179; nectar 178; obscure diets 178; Phyllostomidae 178; *Trachops cirrhosus* (Fringe-lipped Bat) 178, 179
fossils, bat 75
Foster, Richard 104, 106

Galápagos, Ecuador 181–91; *Anoura cultrata* (Handley's Tailless Bat) 186; *Anoura peruana* (Peruvian Tailless Bat) 186, 190; *Anoura* species 201; *Artibeus jamaicensis* 200; *Bassaricyon neblina* (Olinguito) 187; Black Grouse 187; Blue Boobies 182; *Centronycteris centralis* (Shaggy Bat) 198; *Centurio senex* (Wrinkle-faced Bat) 187, 200; cloud forest 188; Cock-of-the-Rock (COTR) 187, 189; *Cormura brevirostris* (Chestnut Sac-winged Bat) 199; *Cyttarops alecto* (Short-eared Bat) 199; Darwin finches 181; *Dermanura tolteca* (Toltec Fruit-eating Bat) 200; *Ectophylla alba* (Honduran White Bat) 198; Elfin forests 189; *Eptesicus andinus* (Little Black Serotine) 186; Galápagos bat species 182; Gálapagos Flycatcher 183; Galápagos Petrel 183; Galápagos Shearwaters 182; Greater Sage-grouse 187; Grey Four-eyed Opossum 200; *Histiotus montanus* (Small Big-eared Brown Bat) 186; *Lasiurus/Aorestes villosissimus* (Hoary Bat) 183; *Lasiurus brachyotis* (Galápagos Yellow Bat) 183; Least Sandpiper 184; *Lophostoma brasiliense* (Pygmy Round-eared Bat) 198; *Lophostoma sylvicolum* (White-throated Round-eared Bat) 198; mist-netting 189; *Myotis oxyotus* (Montane Myotis) 191; *Nyctinomops aurispinosus* (Peale's Free-tailed Bat) 187; oilbirds 188; *Paraponera clavata* (Bullet Ants) 199; *Phoneutria boliviensis* (Brazilian Wandering Spider) 199; Phyllostomidae 186; Pink-footed Shearwater 183; *Porthidium nasutum* (Hognosed Pit-viper) 199; red-billed Tropicbird 184; red-footed Booby 183; Red-necked Phalarope 183; San Cristóbal Mockingbird 183; Semipalmated Plover 184; Sidney Funnel-web Spider 199; Small Tree-finch 183; Sooty Shearwater 183; Storm-petrel 184; *Sturnira erythromus* (Hairy Yellow-shouldered Bat) 189; *Sturnira* species 201; Tayra, Andean White-eared Opossums 187; Waved Albatross 183; Wedge-rumped Storm-petrel 183; Woodpecker Finch 183
Gem-faced Civets, Taiwan 106
geothermal energy 121
ghost bats 14; Darwin, Australia *44*
giant flying squirrels, Sepilok, Borneo 71, 74
giant honeyeater, Viti Levu, Fiji 18
gibbon, Taman Negara National Park, Malaysia 78
Gillanders, Alan 40, 42
Glischropus tylopus (Common Thick-thumbed Bat), Taman Negara National Park, Malaysia 78
Golden Bowerbird hide, Darwin, Australia 45
Golden Dove, Viti Levu, Fiji 20
Golden-rumped Elephant-shrews, Nairobi, Kenya 149
Gomantong Cave, Sepilok, Borneo 72
Goodman's Mouse-lemur, Andasibe, Madagascar 145
Gorecki, Vanessa 34, 36
Grande Terre, New Caledonia 30–33; landscape *31*; *Notopteris neocaledonicus* (New Caledonia Long-tailed Fruit Bat) *32*, 33; *Nyctophilus nebulosus* (New Caledonia Long-eared Bat) 31, 32; *Pteropus ornatus* (Ornate Flying Fox) 33; *Pteropus vetulus* (New Caledonia Flying Fox) 33

Granger, Matt 109
Greater Crested Terns, Viti Levu, Fiji 19
Great Horned Owl, Jalisco, Mexico 126
Green-backed Kingfisher, Tangkoko, Indonesia 57
Green Ibis, Puerto Maldonado, Peru 175
Green Kingfisher, Puerto Maldonado, Peru 175
Grey Four-eyed Opossum, Galápagos, Ecuador 200
Grey-headed Woodpecker, Kaeng Krachan, Thailand 82
Guadalcanal, Solomon Islands 23–27; disturbed habitats for bats 26; *Dobsonia inermis* (Naked-backed Fruit Bat) 26–27; *Macroglossus minimus* (Lesser Long-tongued Blossom Bat) 25; *Miniopterus tristis* (Great Bent-winged Bat) 24; *Mosia nigrescens* (Dark Sheath-tailed Bat) 24; *Myotis moluccarum* (Maluku Myotis) 24; *Myotis myotis* 25; nectar-feeding bat 25; outdoor World War II museum *27*; *Pipistrellus angulatus* (New Guinea Pipistrelle) 24; *Pteropus admiralitatum* (Admiralty Flying Fox) 25; *Saccolaimus saccolaimus* (Bare-rumped Sheath-tailed Bat) 24
Guanay Cormorants, Puerto Maldonado, Peru 171

Halloween decorations 111
Hipposideridae (Old World leaf-nosed bats) 34; St Lucia, South Africa 156
Hipposideros armiger (Greater Himalayan Leaf-nosed Bat): Kaeng Krachan, Thailand 84; Taman Negara National Park, Malaysia 79
Hipposideros ater (Dusky Leaf-nosed Bat), Queensland, Australia 40
Hipposideros diadema (Diadem Leaf-nosed Bat), Queensland, Australia 42
Hipposideros speoris (Leaf-nosed Bat), Mumbai, India 92
Histiotus montanus (Small Big-eared Brown Bat), Galápagos, Ecuador 186
Histiotus velatus (Tropical Big-eared Brown Bat) Galápagos, Ecuador 186
Hogan, Luke 34–36, 124
hydroelectricity 121
Hypsignathus monstrosus (Hammerhead Bat) 87

Icaronycteris index 75, 76
Impalas, St Lucia, South Africa 154
International Bat Night 70, 204–6
invasive species 11; Cane Toads and Ghost Bats 14; Raccoons, Europe 13; rats and stoats, New Zealand 14
Island bats 28–29; Christmas Island 29; *Coleura seychellensis* (Seychelles) 29; conservation shortcomings 28–29; harsh weather 28; *Mirimiri acrodonta* (Fiji) 29; *Mystacina robusta* (Christmas Island) 29; *Mystacina tuberculata* (Lesser Short-tailed Bat) 29; Mystacinidae (New Zealand) 28; Myzopodidae (Madagascar) 28; *Pteralopex/Mirimiri* (Solomon Islands and Fiji) 28; *Pteropus livingstonii* (Comores) 29; translocation attempts 29

Jalisco, Mexico 126–28; *Balantiopteryx plicata* (Gray Sac-winged Bat) 127; Buff-collared Nightjar 126; *Desmodus rotundus* (Vampire Bat) 128; Great Horned Owl 126; Laguna Zapotlán Lake 127; Mexican Whip-poor-will 126; *Myotis yumanensis* (Yuma Myotis) 127; Neotropics 128; *Pteronotus* and *Natalus* species 128; Red Warbler 126; Tiger Heron 127

Kaeng Krachan, Thailand 81–85; *Aselliscus stoliczkanus* (Stoliczka's Trident Bat) 85; *Craseonycteris thonglongyai* (Bumblebee Bat) 84; grey-headed woodpecker 82; *Hipposideros armiger* (Greater Himalayan Leaf-nosed Bat) 84; *Kerivoula picta* (Painted Bat) 82; large-tailed and great-eared nightjars 81; *Miniopterus fuliginosus* (Asian Long-fingered Bat) 81; painted bat habitat 83; 'Painted Bat' village in Eastern Thailand 82; *Rhinolophus lepidus* (Blyth's Horseshoe Bat) 81; *Rhinolophus malayanus* (Malayan Horseshoe Bat) 81; *Tylonycteris* genus 82; *Tylonycteris malayana* (Malayan Bamboo Bat) 81; Yellow-browed Warbler 82
kangaroos or wallaroos, Queensland, Australia 39–40
Kaur, Harpeet 88
Kerivoula picta (Painted Bat): Austin, USA 118; Kaeng Krachan, Thailand 82
Ketola, Chris 171
Kidney Fern 8
'Kokoda' 17
KwaZulu-Natal Bat Appreciation Group, St Lucia, South Africa 155

Kyonemichthys rumengani (Lembeh Pygmy Seadragon), Sulawesi, Indonesia 54

Laguna Zapotlán Lake, Jalisco, Mexico 127
Large-tailed and Great-eared Nightjars, Kaeng Krachan, Thailand 81
Lasiurus/Aorestes villosissimus (Hoary Bat), Galápagos, Ecuador 183
Lasiurus borealis (Eastern Red Bat) 87; Austin, USA 118
Lavia frons (Yellow-winged Bat), Nairobi, Kenya 146–48
Least Sandpiper, Galápagos, Ecuador 184
Lilac-breasted Kingfisher, Tangkoko, Indonesia 57
Lophostoma brasiliense (Pygmy Round-eared Bat), Galápagos, Ecuador 198
Lophostoma sylvicolum (White-throated Round-eared Bat), Galápagos, Ecuador 198
Lord Howe Island 29
The Lord of the Rings 7
Lyroderma lyra (Greater Asian False-vampire Bat), Bengaluru, India 89

Maclean, Jenny 40, 42
macrobats 34
Macroderma gigas (Australian Ghost Bat): Darwin, Australia 44; Yucatán, Mexico 134
Macroglossus minimus (Lesser Long-tongued Blossom Bat), Guadalcanal, Solomon Islands 25
Macronycteris gigas (Giant Leaf-nosed Bat), Nairobi, Kenya 149
Macrophyllum macrophyllum (Long-legged Bat) 47
Macrotus waterhousii (Waterhouse's Leaf-nosed Bat), Sinaloa, Mexico 124
Mahé, Seychelles 160–62; Cape Ternay area 161; *Coleura seychellensis* (Seychelles Sheath-tailed Bat) 160, 161; Crab Plover 160; information on *Coleura* 160; *Pteropus seychellensis* (Seychelles Flying Fox) 160; *Pteropus seychellensis* (Seychelles Fruit Bat) 162; Seychelles Paradise Flycatcher 160; Seychelles Parrot 160; Seychelles Scops-owl 160
Mareeba Rock-wallabies, Queensland, Australia 38
Master and Commander (2004) 181
Medellín, Rodrigo 134, 135
Megaderma spasma (Lesser Asian False-vampire Bat), Tangkoko, Indonesia 57, *58*
Merlin Tuttle Bat Conservation 202; Austin, USA 116
MERS 98
Metasepia pfefferi (Flamboyant Cuttlefish), Sulawesi, Indonesia 54, 56
Mexican Whip-poor-will, Jalisco, Mexico 126
Microbat bat detector, Stag Electronics 11
Micronycteris microtis (Common Big-eared Bat), Yucatán, Mexico 130
migration, bat 169–70; *Eidolon helvum* (Straw-coloured Fruit Bat) 169–70; *Pipistrellus nathusii* (Nathusius's Pipistrelle) 169; satellite tracking 170
Mikado Pheasant, Taiwan 104, 105
Miles, Damian 43, 44
Mimon cozumelae (Cozumelan Golden Bat), Yucatán, Mexico 131, 132
Miniopterus australis (Little Long-fingered Bat), Queensland, Australia 36
Miniopterus fraterculus (Lesser Long-fingered Bat), St Lucia, South Africa 155
Miniopterus fuliginosus (Asian Long-fingered Bat), Kaeng Krachan, Thailand 81
Miniopterus fuliginosus (Eastern Bent-winged Bat), Okinawa, Japan 108
Miniopterus fuscus (Southeast Asian Long-fingered Bat), Okinawa, Japan 108
Miniopterus inflatus (Greater Long-fingered Bat), St Lucia, South Africa 156
Miniopterus majori (Major's Long-fingered Bat), Andasibe, Madagascar 143
Miniopterus medius (Intermediate Long-fingered Bat), Taman Negara National Park, Malaysia 79
Miniopterus natalensis (Natal Long-fingered Bat) St Lucia, South Africa 156
Miniopterus orianae (Australasian Long-fingered Bat), Queensland, Australia 34
Miniopterus schreibersii (Common Bent-wing Bat) 158
Miniopterus species, Andasibe, Madagascar 142
Miniopterus tristis (Great Bent-winged Bat), Guadalcanal, Solomon Islands 24
Mirimiri acrodonta (Fijian Monkey-faced Bat), Fiji 29
Mist-netting, Galápagos, Ecuador 189
Molossids, Yucatán, Mexico 135–36

Molossus alvarezi (Alvarez's Mastiff Bat), Yucatán, Mexico 133
Molossus pretiosus (Mastiff Bat), Rio Claro, Costa Rica 195
Molossus rufus (Black Mastiff Bat), Rio Claro, Costa Rica 195; Yucatán, Mexico 130
Monarch Butterfly 169
Mops leucostigma (Malagasy White-bellied Free-tailed Bat), Antsiranana, Madagascar 139
Mormoops megalophylla (Ghost-faced Bat), Yucatán, Mexico 133
Mosia nigrescens (Dark Sheath-tailed Bat), Guadalcanal, Solomon Islands 24; Tangkoko, Indonesia 62; West Papua, Indonesia 63
Mount Mahawu 59
Mumbai, India 91–94; Elephanta Caves 92; *Hipposideros speoris* (Schneider's Leaf-nosed Bat) 92; *Pipistrellus ceylonicus* (Kelaart's Pipistrelle) 93; *Rhinolophus rouxii* (Rufous Horseshoe Bat) 93; Ring-necked Parakeets 92; Sanjay Gandhi National Park 92; *Taphozous melanopogon* (Black-bearded Tomb Bat) 93
Myotis blythii (Lesser Mouse-eared Bat), Chengdu, China 96
Myotis bocagii (Bocage's Myotis), St Lucia, South Africa 154
Myotis brandtii (Brandt's Bat) 86
Myotis browni, Tangkoko, Indonesia 59
Myotis chinensis (Large Myotis), Chengdu, China 96
Myotis crypticus (Cryptic Myotis) 158
Myotis dasycneme (Pond Bat): Chengdu, China 95; Queensland, Australia 36
Myotis daubentonii (Daubenton's Bat), Queensland, Australia 36
Myotis emarginatus (Geoffroy's Bat) 112
Myotis escalerai (Escalera's Bat) 158
Myotis goudotii (Malagasy Myotis), Antsiranana, Madagascar 141
Myotis hasseltii (Hasselt's Myotis), West Papua, Indonesia 68
Myotis lucifugus (Little Brown Bat), Austin, USA 118
Myotis macropus (Large-footed Bat), Queensland, Australia 34, 36, 40
Myotis moluccarum (Maluku Myotis): Guadalcanal, Solomon Islands 24; West Papua, Indonesia 64
Myotis muricola (Nepalese Whiskered Bat), Tangkoko, Indonesia 58
Myotis myotis (Greater Mouse-eared Bat), Chengdu, China 96; Guadalcanal, Solomon Islands 25
Myotis nattereri (Natterer's Bat) 158; Virelles, Belgium 115
Myotis oxyotus (Montane Myotis): Galápagos, Ecuador 191; Yucatán, Mexico 130
Myotis pilosatibialis (Hairy-legged Myotis), Yucatán, Mexico 130
Myotis tricolor (Temminck's Myotis), St Lucia, South Africa 156
Myotis velifer (Cave Myotis), Austin, USA 118
Myotis vivesi (Mexican Fishing Bat), Queensland, Australia 36
Myotis yumanensis (Yuma Myotis), Jalisco, Mexico 127
Mystacina robusta (Christmas Island), New Zealand 29
Mystacina tuberculata (Short-tailed Bat), New Zealand 8, 12, 29
Mystacinidae, New Zealand 28
Myzopodidae, Madagascar 28

Nairobi, Kenya 146–51; *Cardioderma cor* (Heart-nosed Bat) 147–48; caves filled with millions of bats 150; *Coleura afra* (African Sheath-tailed Bat) 149; extracting bat from mist-net 150; Golden-rumped Elephant-shrews 149; *Lavia frons* (Yellow-winged Bat) 146–48; *Macronycteris gigas* (Giant Leaf-nosed Bat) 149; *Nycteris thebaica* (Egyptian Slit-faced Bat) 151; Samburu in central Kenya 146; Samburu National Reserve 146; *Scotophilus andrewreborii* (Yellow Bat) 151; *Scotophilus trujilloi* (Yellow Bat) 151; *Triaenops afer* (African Trident Bat) 149
Narina Trogon, St Lucia, South Africa 153
Natagora-Jeunes (NJ), Virelles, Belgium 112
Natalus species, Jalisco, Mexico 128
nectar-feeding bat 25
Negev desert, Tel Aviv, Israel 163
Neoromicia matroka (Malagasy Serotine), Andasibe, Madagascar 143
Neotropical molossids, Rio Claro, Costa Rica 195
Neotropics, Jalisco, Mexico 128
New Zealand 7–9; beech forests 7; Blue Duck/Whio 8; *Chalinolobus tuberculatus* (Long-tailed Bat) 8, 9; Eglinton Valley 7; Fiordland 8; Kaka 8;

Mystacina 8; *Mystacina tuberculata* (Short-tailed Bat) 8; Pekapeka 8; Pipipi 8; Pureora Forest Park 9; Rock Wren 7, 8; sanctuary islands, Ulva and Blumine/Oruawairua 8; Southern Brown Kiwi/Tokoeka 8; urban penguins, Oamaru 7; Yellowhead/Mohua 8
Nichta, Teresa 117, 119
Niviventer rats, Taiwan 106
Nixon, David 43, 44
Noctilio albiventris (Lesser Bulldog Bat), Puerto Maldonado, Peru 174, 176
Noctilio leporinus (Greater Bulldog Bat), Yucatán, Mexico 136
Northern Rufous Mouse-lemur, Antsiranana, Madagascar 140
Notopteris macdonaldii (Long-tailed Fruit Bat), Viti Levu, Fiji 20, 22
Nyctalus lasiopterus (Giant Noctule), Eilat, Israel 167
Nyctalus noctula (Noctule Bat) 87
Nycteris thebaica (Egyptian Slit-faced Bat), Nairobi, Kenya 151
Nycteris tragata (Malayan Slit-faced Bat), West Papua, Indonesia 68
Nycticeius humeralis (Evening Bat), Austin, USA 119
Nyctimene cephalotes (Pallas's Tube-nosed fruit Bat), Tangkoko, Indonesia 61
Nyctimene robinsoni (Queensland Tube-nosed fruit Bat), Queensland, Australia 38
Nyctinomops aurispinosus (Peale's Free-tailed Bat), Galápagos, Ecuador 187
Nyctinomops laticaudatus (Broad-eared Free-tailed Bat), Yucatán, Mexico 129
Nyctophilus bifax (Eastern Long-eared Bat), Queensland, Australia *41*, 41–42
Nyctophilus gouldi (Gould's Long-eared Bat), Queensland, Australia 42
Nyctophilus nebulosus (New Caledonia Long-eared Bat), Grande Terre, New Caledonia 31, 32

oilbirds, Galápagos, Ecuador 188
Okinawa, Japan 108–9; Hokkaido 108–9; *Miniopterus fuliginosus* (Eastern Bent-winged Bat) 108; *Miniopterus fuscus* (Southeast Asian Long-fingered Bat) 108; *Rhinolophus pumilus* (Little Okinawan horseshoe Bat) 108; *Rhinolophus pusillus* (Least Horseshoe Bat) 108
Old World tarantulas 59
orange-haired wingless bat, Sepilok, Borneo 73
Orangutan Crab, Sulawesi, Indonesia 56
Otomops wroughtoni (Wroughton's Giant Mastiff Bat), Bengaluru, India 88
Otonycteris hemprichii (Desert Long-eared Bat), Tel Aviv, Israel 163, 164

Pacific Golden Plover 19
painted bat habitat, Kaeng Krachan, Thailand 83
Palawan Peacock-pheasants, Subic Bay, Luzon, Philippines 50
Palawan Scops-owl, Subic Bay, Luzon, Philippines 51
Panda 96
Paraemballonura tiavato (Western Sheath-tailed Bat), Antsiranana, Madagascar 140
Paraponera clavata (Bullet Ants), Galápagos, Ecuador 199
Parque Nacional Carara, Rio Claro, Costa Rica 192
Parsons, Stuart 45, 162
Pekapeka 8
Pennay, Michael 24
Peruvian Boobies, Puerto Maldonado, Peru 171
Peruvian Pelicans, Puerto Maldonado, Peru 171
Philippine Frogmouth, Subic Bay, Luzon, Philippines *51*
Philippine Slow Loris, Sepilok, Borneo 72
Phoneutria boliviensis (Brazilian Wandering Spider), Galápagos, Ecuador 199
Phyllostomidae: Galápagos, Ecuador 186; Puerto Maldonado, Peru 174, 177; Rio Claro, Costa Rica 194
Phyllostomidae (New World leaf-nosed bats) 34
Phyllostomus hastatus (Greater Spear-nosed Bat), Puerto Maldonado, Peru 171–72, 177
Pink-footed Shearwater, Galápagos, Ecuador 183
Pipipi 8
Pipistrellus abramus (Japanese Pipistrelle) 4
Pipistrellus angulatus (New Guinea Pipistrelle), Guadalcanal, Solomon Islands 24
Pipistrellus ceylonicus (Pipistrelle), Mumbai, India 93
Pipistrellus collinus (Greater Papuan Pipistrelle), West Papua, Indonesia 66
Pipistrellus javanicus (Javan Pipistrelle), Subic Bay, Luzon, Philippines 49
Pipistrellus kuhlii (Kuhl's Pipistrelle), Tel Aviv, Israel 163

Pipistrellus maderensis (Madeiran Pipistrelle), Eilat, Israel 167
Pipistrellus nathusii (Nathusius's Pipistrelle) 169
Pipistrellus rusticus (Rusty Pipistrelle), St Lucia, South Africa 156, 157
Pipistrellus stenopterus (Narrow-winged Pipistrelle), Taman Negara National Park, Malaysia *79*
Platyrrhinus helleri (Heller's Broad-nosed Bat), Rio Claro, Costa Rica 196
Platyrrhinus infuscus (Buffy Broad-nosed Bat), Puerto Maldonado, Peru 175
Plecotus christii (Christie's Long-eared Bat), Eilat, Israel 165, 167
Plecotus taivanus (Taiwan Long-eared Bat), Taiwan *79*, 106
Plecotus teneriffae (Tenerife Long-eared Bat): Eilat, Israel 167; Tenerife, Canary Islands 3
polyestrous species 87
Porthidium nasutum (Hognosed Pit-viper), Galápagos, Ecuador 199
'Prairies bocagères,' Virelles, Belgium 112
Prionoplus reticularis (Huhu beetles), Pureora, North Island, New Zealand 11
Promops centralis (Big Crested Mastiff Bat), Yucatán, Mexico 136
Promops nasutus (Brown Mastiff Bat), Puerto Maldonado, Peru 173
Pteronotus fulvus (Tawny Naked-backed Bat), Yucatán, Mexico 133
Pteronotus mesoamericanus (Mesoamerican Moustached Bat), Yucatán, Mexico 130, 133
Pteropodidae 47
Pteropus admiralitatum (Admiralty Flying Fox), Guadalcanal, Solomon Islands 25
Pteropus alecto (Black Flying Fox), Queensland, Australia 36
Pteropus conspicillatus (Spectacled Flying Fox), Queensland, Australia 36
Pteropus griseus (Grey Flying Fox), Tangkoko, Indonesia 60
Pteropus livingstonii (Livingstone's Fruit Bat) Comores 29
Pteropus macrotis (Big-eared Flying Fox), West Papua, Indonesia 66
Pteropus ocularis (Seram Flying Fox), Tangkoko, Indonesia 62
Pteropus ornatus (Ornate Flying Fox), Grande Terre, New Caledonia 33
Pteropus samoensis (Samoa Flying Fox), Viti Levu, Fiji 20
Pteropus seychellensis (Seychelles Flying Fox) 160; Seychelles Fruit Bat 162
Pteropus temminckii (Temminck's Flying Fox), Tangkoko, Indonesia 61, *62*
Pteropus tonganus (Pacific Flying Fox), Viti Levu, Fiji *21*
Pteropus vampyrus (Large Flying Fox): Subic Bay, Luzon, Philippines 49, 50; West Papua *67*, 68
Pteropus vetulus (New Caledonia Flying Fox), Grande Terre, New Caledonia 33
Puerto Maldonado, Peru 171–77; birdwatching 171; *Carollia perspicillata* (Short-tailed Bat) 174; *Chironius fuscus* (Brown Sipo) 172; Green Ibis 175; Green Kingfisher 175; Guanay Cormorants 171; *Molossus* species 171; *Noctilio albiventris* (Lesser Bulldog Bat) 174, 176; Peruvian Boobies 171; Peruvian Pelicans 171; Phyllostomidae 174, 177; *Phyllostomus* and *Carollias* 177; *Phyllostomus hastatus* (Greater Spear-nosed Bat) 171–72; *Platyrrhinus infuscus* (Buffy Broad-nosed Bat) 175; *Promops nasutus* (Brown Mastiff Bat) 173; *Rhinophylla pumilio* (Dwarf Little Fruit Bat) 177; *Rhynconycteris naso* (Proboscis Bat) 173; *Saccopteryx bilineata* (Greater Sac-winged Bat) 173; Spectacled Owl 176; Tambopata River 173, 175; *Thyroptera tricolor* (Spix's Disk-winged Bat) 176; *Trachops cirrhosus* (Fringe-lipped Bat) 172; Western Striolated Puffbird 175; Wilson's and Elliot's Storm-petrels 171; Yellow-footed Tortoise 173
Puerto Princesa Underground River, Subic Bay, Luzon, Philippines 51, *51*
Pureora, North Island, New Zealand 10–12; catching bats with department of conservation *11*; *Chalinolobus* 11, 12; harp traps, benefits 10–11; Microbat bat detector, Stag Electronics 11; *Mystacina tuberculata* (Short-tailed Bat) 12; *Prionoplus reticularis* (Huhu beetles) 11
Pureora Forest Park 9
Puttaswamaiah, Rajesh 88–91

Queensland, Australia 34–42; arboreal marsupials 42; *Austronomus australis* (White-striped Free-tailed Bat) 36; *Chaerephon jobensis* (Greater Northern Free-tailed Bat) 36; echidnas 37; echolocating bats 34; *Hipposideros ater* (Dusky Leaf-nosed Bat) 40; *Hipposideros diadema* (Diadem

Leaf-nosed Bat) 42; kangaroos or wallaroos 39–40; Mareeba rock-wallabies 38; *Miniopterus australis* (Little Long-fingered Bat) 36; *Miniopterus orianae* (Australasian Long-fingered Bat) 34; *Myotis dasycneme* (Pond Bat) 36; *Myotis daubentonii* (Daubenton's Bat) 36; *Myotis macropus* (Large-footed Bat) 34, 36, 40; *Myotis vivesi* (Mexican Fishing Bat) 36; *Nyctimene robinsoni* (Tube-nosed Fruit Bat) 38; *Nyctophilus bifax* (Eastern Long-eared Bat) *41*, 41–42; *Nyctophilus gouldi* (Long-eared Bat) 42; *Pteropus alecto* (Black Flying Fox) 36; *Pteropus conspicillatus* (Spectacled Flying Fox) 36; rock-wallabies 39, *39*; *Saccolaimus flaviventris* (Yellow-bellied Sheath-tailed Bat) 36; *Scotorepens* sp. (Broad-nosed Bat) 35; Striped Possum 42; *Taphozous australis* (Coastal Sheath-tailed Bat) 36; urban flying foxes, Brisbane *35*; *Vespadelus pumilus* (Eastern forest Bat) 42; Wet Tropics 37

'quiet' bats 47

Quinnell, Abi 10–12

Quokkas from Rottnest Island 38

raccoons, Europe 13

Racey, Paul 161

rats, New Zealand 14

Red-backed Shrike, Virelles, Belgium 112

Red-billed Tropicbird, Galápagos, Ecuador 184

Red-footed Booby, Galápagos, Ecuador 183

Red-necked Phalarope, Galápagos, Ecuador 183

Red Warbler, Jalisco, Mexico 126

reproduction 86–87; bat pup 86; *Desmodus rotundus* (Common Vampire Bat) 86; gestation period 86–87; *Hypsignathus monstrosus* (Hammerhead Bat) 87; *Lasiurus borealis* (Eastern red Bat) 87; life expectancy 86; *Myotis brandtii* (Brandt's Bat) 86; *Nyctalus noctula* (Noctule Bat) 87; polyestrous species 87; sexual dimorphism 87; *Sphaeronycteris toxophyllum* (Visored Bat) 87

Rhinolophidae (Horseshoe Bats) 34

Rhinolophus beddomei (Horseshoe Bat), Bengaluru, India 88–90

Rhinolophus celebensis (Sulawesi Horseshoe Bat), Tangkoko, Indonesia 59

Rhinolophus clivosus (Geoffroy's Horseshoe Bat), St Lucia, South Africa 156

Rhinolophus euryotis (Broad-eared Horseshoe Bat), West Papua, Indonesia 63–64

Rhinolophus formosae (Formosan woolly horseshoe Bat), Taiwan 101–2

Rhinolophus lepidus (Blyth's Horseshoe Bat), Kaeng Krachan, Thailand 81

Rhinolophus malayanus (Malayan horseshoe Bat), Kaeng Krachan, Thailand 81

Rhinolophus monoceros (Formosan lesser horseshoe Bat), Taiwan 101

Rhinolophus philippinensis (Large-eared Horseshoe Bat): Subic Bay, Luzon, Philippines 49; West Papua, Indonesia 66

Rhinolophus pumilus (Little Okinawan Horseshoe Bat), Okinawa, Japan 108

Rhinolophus pusillus (Least Horseshoe Bat), Okinawa, Japan 108

Rhinolophus rouxii (Rufous horseshoe Bat), Mumbai, India 93

Rhinolophus simulator (Bushveld Horseshoe Bat), St Lucia, South Africa 155

Rhinolophus trifoliatus (Trefoil horseshoe Bat), Sepilok, Borneo 74

Rhinonicteris aurantia (Orange Leaf-nosed Bat): Darwin, Australia 44; Queensland, Australia 40

Rhinophylla pumilio (Dwarf Little Fruit Bat), Puerto Maldonado, Peru 177

Rhinopoma cystops (Big-eyed Mouse-tailed Bat), Eilat, Israel 165, 167

Rhogeessa tumida (Black-winged Little Yellow Bat), Yucatán, Mexico 133

Rhynconycteris naso (Proboscis Bat): Puerto Maldonado, Peru 173; Rio Claro, Costa Rica 193

Ring-necked Parakeets, Mumbai, India 92

Ring-tailed Lemurs, Andasibe, Madagascar 142

Rio Claro, Costa Rica 192–201; Bat Blitz 192; Black Mastiff Bat 195; *Dermanura watsoni* (Thomas's fruit-eating Bat) 197; *Desmodus rotundus* (Common Vampire Bat) 195; *Diclidurus albus* (Northern ghost Bat) 192, 193; *Enchistenes hartii* (Velvety Fruit-eating Bat) 196; *Molossus pretiosus* (Mastiff Bat) 195; *Molossus rufus* (Black Mastiff Bat) 195; *Molossus* species 194;

Neotropical molossids 195; Parque Nacional Carara 192; Phyllostomidae 194; *Platyrrhinus helleri* (Heller's Broad-nosed Bat) 196; *Rhynconycteris naso* (Proboscis Bat) 193; *Saccopteryx bilineata* (Greater Sac-winged Bat) 193; Scarlet Macaws 193; *Thyroptera discifera* (Disk-winged Bat) 196; *Vampyressa thyone* (Northern Little yellow-eared Bat) 196
Rock Wren 7, 8
Round the World Flights, British travel agency 5
Rousettus aegyptiacus (Egyptian fruit Bat), Tel Aviv, Israel 163
Rousettus amplexicaudatus (Geoffroy's fruit Bat), Subic Bay, Luzon, Philippines 51
Ruby-throated Hummingbird, Austin, USA 119

Saccolaimus flaviventris (Yellow-bellied Sheath-tailed Bat), Queensland, Australia 36
Saccolaimus saccolaimus (Bare-rumped Sheath-tailed Bat), Guadalcanal, Solomon Islands 24
Saccopteryx bilineata (Greater Sac-winged Bat): Puerto Maldonado, Peru 173; Rio Claro, Costa Rica 193
Sambar Deer, Taman Negara National Park, Malaysia 78
Samburu National Reserve, Nairobi, Kenya 146
San Cristóbal Mockingbird, Galápagos, Ecuador 183
Sanjay Gandhi National Park, Mumbai, India 92
SARS 98
SARS-CoV-2 98
Scarlet Macaws, Rio Claro, Costa Rica 193
Scotophilus andrewreborii (Yellow Bat), Nairobi, Kenya 151
Scotophilus dinganii (African yellow Bat), St Lucia, South Africa 154
Scotophilus heathii (Greater Asiatic yellow house Bat), Taman Negara National Park, Malaysia 78
Scotophilus robustus (Robust House Bat), Andasibe, Madagascar 145
Scotophilus species, Andasibe, Madagascar 142
Scotophilus trujilloi (Yellow Bat), Nairobi, Kenya 151
Scotophilus viridis (Eastern greenish yellow Bat), St Lucia, South Africa 154
Scotorepens sp. (Broad-nosed Bat), Queensland, Australia 35
Semipalmated Plover, Galápagos, Ecuador 184
Sepilok, Borneo 71–74; *Cheiromeles torquatus* (Naked Bat) 71; *Cyrtodactylus consobrinus* (Banded Forest Gecko) 73; giant flying squirrels 71, 74; Gomantong Cave 72; orange-haired wingless bat 73; Philippine Slow Loris 72; *Rhinolophus trifoliatus* (Trefoil horseshoe Bat) 74; sleeping Rufous-backed Dwarf Kingfisher 72; slow lorises 73; *Thelyphonida* 73; *Tropidolaemus wagleri* (Wagler's pit viper) 71; whip spiders (*Amblypygi*) 73
sexual dimorphism 87
Seychelles (*Coleura seychellensis*) 29
Short-tailed Brushtail Possum, Darwin, Australia 45
Sidney Funnel-web Spider, Galápagos, Ecuador 199
Sinaloa, Mexico 124–25; Culiacan 124; landscape 125; *Macrotus waterhousii* (Waterhouse's Leaf-nosed Bat) 124; Snowy Plover research team 124
Small Tree-finch, Galápagos, Ecuador 183
social distancing 99
solar power 121
Sooty Shearwater, Galápagos, Ecuador 183
Southern Brown Kiwi/Tokoeka 8
Spectacled Owl, Puerto Maldonado, Peru 176
Spectral Tarsier roost, Tangkoko, Indonesia 57
Sphaeronycteris toxophyllum (Visored Bat) 87
Steatornis caripensis (Oilbird) 47
St Lucia, South Africa 146–57; *Cloeotis percivali* (Short-eared Trident Bat) 156; Cradle of Humankind, World Heritage Site 155; crocs and bats 154; *Eptesicus hottentotus* (Hottentot Serotine Bat) 152; Hipposideridae 156; Impalas 154; KwaZulu-Natal Bat Appreciation Group 155; *Miniopterus fraterculus* (Lesser Long-fingered Bat) 155; *Miniopterus inflatus* (Greater Long-fingered Bat) 156; *Miniopterus natalensis* (Natal Long-fingered Bat) 156; *Myotis bocagii* (Bocage's Myotis) 154; *Myotis tricolor* (Temminck's Myotis) 156; Narina Trogon 153; *Pipistrellus rusticus* (Rusty

Pipistrelle) 156, 157; *Rhinolophus clivosus* (Geoffroy's Horseshoe Bat) 156; *Rhinolophus simulator* (Bushveld Horseshoe Bat) 155; *Scotophilus dinganii* (African yellow Bat) 154; *Scotophilus viridis* (Eastern greenish yellow Bat) 154
stoats, New Zealand 14
Striped Possum, Queensland, Australia 42
Sturnira erythromus (Hairy Yellow-shouldered Bat), Galápagos, Ecuador 189
Sturnira species, Galápagos, Ecuador 201
Subic Bay, Luzon, Philippines 49–53; *Acerodon jubatus* (Golden-crowned Flying Fox) 49; *Acerodon leucotis* (Palawan Flying Fox) 51; *Chaerephon plicatus* (Wrinkle-lipped Free-tailed Bat) 49, 53; *Cynopterus luzoniensis* (Peter's fruit Bat) *53*; Palawan, Island 50; Palawan Peacock-pheasants 50; Palawan Scops-owl 51; Philippine Frogmouth *51*; *Pipistrellus javanicus* (Javan Pipistrelle) 49; *Pteropus vampyrus* (Large Flying Fox) 49; Puerto Princesa Underground River 51, *51*; *Rhinolophus philippinensis* (Large-eared Horseshoe Bat) 49; *Rousettus amplexicaudatus* (Geoffroy's fruit Bat) 51
Sulawesi Black Tarantula, Tangkoko, Indonesia 59
Synchiropus splendidus (Mandarin Goby), Sulawesi, Indonesia 54

Tachypteron franzeni 76
Tadarida aegyptiaca (Egyptian Free-tailed Bat), Bengaluru, India 89
Tadarida brasiliensis (Mexican Free-tailed Bat), Austin, USA 118, 119
Tadarida fulminans (Malagasy Free-tailed Bat), Antsiranana, Madagascar 139
Tadarida insignis (East Asian Free-tailed Bat), Taiwan 105
Tadarida teniotis (European Free-tailed Bat): Tel Aviv, Israel 163; Tenerife, Canary Islands 2
Taiwan 101–7; Bamboo-partridge 107; *Barbastella darjelingensis* (Darjeeling Barbastelle) 106; black-necklaced Scimitar-babbler 107; Brown Dipper 107; Bush Warbler 106; Collared Owlet 106; Cupwing 106; Daxueshan National Forest 105; *Eptesicus pachyomus* (Mouse-like Serotine) 104; Flamecrest 106; Gem-faced Civets 106; Laughingthrushes 104; Mikado Pheasant 104, 105; *Niviventer* rats 106; *Plecotus taivanus* (Taiwan Long-eared Bat) 106; *Rhinolophus formosae* (Formosan woolly horseshoe Bat) 101–2; *Rhinolophus monoceros* (Formosan lesser horseshoe Bat) 101; Rosefinch 105; Slug Snake 105; *Tadarida insignis* (East Asian Free-tailed Bat) 105; Taiwanese Red and White Giant Flying Squirrel 106; *Trimeresurus stejnegeri* (Chinese Bamboo pit-vipers) 103; *Vespertilio sinensis* (Asian Particoloured Bat) 105; Yellow Tit 104
Taman Negara National Park, Malaysia 78–80; *Chaerephon johorensis* (Lesser Northern Free-tailed Bat) 79; *Chaerephon plicatus* (Wrinkle-lipped Free-tailed Bat) 78; *Cynopterus brachyotis* (Forest Short-nosed Fruit Bats) 79; gibbon 78; *Glischropus tylopus* (Common Thick-thumbed Bat) 78; *Hipposideros armiger* (Greater Himalayan leaf-nosed Bat) 79; *Miniopterus medius* (Intermediate Long-fingered Bat) 79; *Pipistrellus stenopterus* (Narrow-winged Pipistrelle) 79; Sambar Deer 78; *Scotophilus heathii* (Greater Asiatic yellow house Bat) 78; Wild Boar 78
Tambopata River, Puerto Maldonado, Peru 173, 175
Tangkoko, Indonesia 57–62; *Acerodon celebensis* (Sulawesi Flying Fox) 60, *61*; *Aselliscus tricuspidatus* (Temminck's Trident) 62; *Dobsonia viridis* (Greenish Naked-backed Fruit Bat) 61; documenting bat-hunting 60; 'Extreme' Market of Tomohon 60; Green-backed Kingfisher 57; Isabelline Bush-hen 59; Lilac-breasted Kingfisher 57; *Megaderma spasma* (Lesser Asian False-vampires) 57, *58*; *Mosia nigrescens* (Dark Sheath-tailed Bat) 62; *Myotis browni* 59; *Myotis muricola* (Nepalese whiskered Bat) 58; *Nyctimene cephalotes* (Tube-nosed Fruit Bat) 61; *Pteropus griseus* (Grey Flying Fox) 60; *Pteropus ocularis* (Seram Flying Fox) 62; *Pteropus temminckii* (Temminck's Flying Fox) 61, *62*; *Rhinolophus celebensis* (Sulawesi Horseshoe Bat) 59; Spectral Tarsier roost 57; Sulawesi Black Tarantula 59;

Taphozous melanopogon (Black-bearded Tomb Bat) 58; *Taphozous* species 59
Taphozous australis (Coastal Sheath-tailed Bat), Queensland, Australia 36
Taphozous melanopogon (Black-bearded Tomb Bat): Mumbai, India 93; Tangkoko, Indonesia 58
Taphozous nudiventris (Naked-rumped Tomb Bat), Eilat, Israel 167
Taphozous species: Darwin, Australia 44; Tangkoko, Indonesia 59
Tasmania, Australia 29
taxonomy 158–59; bats, taxonomic golden age 158; importance 158; IUCN Red List 159; *Miniopterus schreibersii* 158; *Myotis crypticus* 158; *Myotis escalerai* 158; *Myotis nattereri* 158
Tayra, Andean White-eared Opossums, Galápagos, Ecuador 187
Tel Aviv, Israel 163–64; *Asellia tridens* (Trident leaf-nosed Bat) 163; Negev desert 163; *Otonycteris hemprichii* (Desert Long-eared Bat) 163, 164; *Pipistrellus kuhlii* (Kuhl's Pipistrelle) 163; *Rousettus aegyptiacus* (Egyptian fruit Bat) 163; *Tadarida teniotis* (European Free-tailed Bat) 163
Tenerife, Canary Islands 1–6; Big Bat Year planning 2; hearing 4; landscape *3*; *Plecotus teneriffae* (Tenerife Long-eared Bat) 3; *Tadarida teniotis* (European Free-tailed Bat) 2
The Thin Red Line 23, 40
Thyroptera discifera (Peter's Disk-winged Bat), Rio Claro, Costa Rica 196
Thyroptera tricolor (Spix's Disk-winged Bat), Puerto Maldonado, Peru 176
Tiger Heron, Jalisco, Mexico 127
Tolga Bat Hospital 38
Trachops cirrhosus (Fringe-lipped Bat) 178; Puerto Maldonado, Peru 172
travelling, importance of 205
Triaenops afer (African Trident Bat), Nairobi, Kenya 149
Triaenops menamena (Rufous Trident Bat), Antsiranana, Madagascar 140, 141
Tropidolaemus wagleri (Wagler's pit viper), Sepilok, Borneo 71
Twelve-wire Bird-of-Paradise, West Papua, Indonesia 67
Tylonycteris genus, Kaeng Krachan, Thailand 82
Tylonycteris malayana (Malayan Bamboo Bat), Kaeng Krachan, Thailand 81

UNESCO World Heritage Site 92
urban flying foxes, Brisbane, Queensland, Australia *35*
urban penguins, Oamaru 7

vampires and people 137–38; *Desmodus rotundus* (Common Vampire Bat) 137; feeding on blood of cows 138; impact on cattle and people 138; rise in rabies cases 138
Vampyressa thyone (Northern Little Yellow-eared Bat), Rio Claro, Costa Rica 196
Vampyrum spectrum (Spectral Bat), Yucatán, Mexico 134
Vespadelus pumilus (Eastern Forest Bat), Queensland, Australia 42
Vespertilio murinus (Particoloured Bat), Eilat, Israel 167
Vespertilionidae (Vesper Bats) 34
Vespertilio sinensis (Asian Particoloured Bat), Taiwan 105
Virelles, Belgium 112–15; Aquatic Warblers 114; Barbastelles 115; bats in exotic places 113; Biebrza National Park 115; Chimay region 112; Collared Flycatchers 114; Geoffroy's Bat 112; Great Crested Newt 112; grey-headed woodpecker 115; Lesser and Greater Horseshoe Bat 112; *Myotis nattereri* (Natterer's Bat) 115; Natagora-Jeunes (NJ) 112; 'Prairies bocagères' 112; red-backed Shrike 112; Serotine *Eptesicus serotinus* 115; Southern Damselfly 112
Viti Levu, Fiji 17–22; caves, role of 21–22; Colo-I-Suva Rainforest Eco Resort 17; *Emballonura semicaudata* (Pacific Sheath-tailed Bat) 21, 22; Fiji Bush-warbler 20; giant honeyeater 18; golden dove 20; greater crested terns 19; landscape *19*; *Notopteris macdonaldii* (Long-tailed Fruit Bat) 20, 22; Pacific Golden Plover 19; *Pteropus samoensis* (Samoa Flying Fox) 20; *Pteropus tonganus* (Pacific Flying Fox) *21*; Suva 17–18; uniqueness 20; Wandering Tattler 19
volcano of bats, Yucatán, Mexico 133
von Nathusius, Hermann 169

Waigeo, West Papua, Indonesia 64–65
Waldien, Dave 21, 22
Waved Albatross, Galápagos, Ecuador 183
Wedge-rumped Storm-petrel, Galápagos, Ecuador 183

Western Striolated Puffbird, Puerto Maldonado, Peru 175
Western Yellow Bat (*Dasypterus xanthinus*), Austin, USA 119
West Papua, Indonesia 63–68; *Aselliscus tricuspidatus* (Temminck's Trident Bat) 64; cloud forest of Arfak Mountains 66; *Dobsonia* sp. (Naked-backed Fruit Bat) 65; echolocation data 66; *Mosia nigrescens* 63; *Myotis hasseltii* (Hasselt's Myotis) 68; *Myotis moluccarum* (Maluku Myotis) 64; *Nycteris tragata* (Malayan Slit-faced Bat) 68; *Pipistrellus collinus* (Greater Papuan Pipistrelle) 66; *Pteropus macrotis* (Big-eared Flying Fox) 66; *Pteropus vampyrus* (Large Flying Fox) 67, 68; *Rhinolophus euryotis* (Broad-eared Horseshoe Bat) 63–65; *Rhinolophus philippinensis* (Large-eared Horseshoe Bat) 66; Twelve-wire Bird-of-Paradise 67; Waigeo 64–65
whip spiders (*Amblypygi*), Sepilok, Borneo 73
Wild Boar, Taman Negara National Park, Malaysia 78
Wildlife Acoustics EM3+, 114
Wilson's and Elliot's Storm-petrels, Puerto Maldonado, Peru 171
wind turbines 121–22; acoustic jamming 122; collisions for bats 121; decline in bat population 121; passive and active solutions 122; passive solutions 122; Ptarmigan mortality 122
Woodpecker Finch, Galápagos, Ecuador 183

Yellow-browed Warbler, Kaeng Krachan, Thailand 82
Yellow-footed Tortoise, Puerto Maldonado, Peru 173
Yellowhead/Mohua 8
Yucatán, Mexico 129–36; *Artibeus jamaicensis* (Jamaican Fruit-eating Bat) 129; *Carollia perspicillata* (Short-tailed Bat) 130; *Carollia sowelli* (Sowell's Short-tailed Bat) 130; *Chrotopterus auritus* (Woolly False-vampire Bat) 134; *Desmodus rotundus* (Common Vampire Bat) 131; *Diphylla ecaudata* (Hairy-legged Vampire Bat) 132; *Eptesicus furinalis* (Argentinian Serotine) 131; *Eumops ferox* (Wild Bonneted Bat) 130; *Eumops underwoodi* (Underwood's Bonneted Bat) 133; *Macroderma gigas* (Australian Ghost Bat) 134; Mayan temples, home to many bats 134; *Micronycteris microtis* (Common Big-eared Bat) 130; *Mimon cozumelae* (Cozumelan Golden Bat) 131, 132; molossids 135–36; *Molossus alvarezi* (Alvarez's Mastiff Bat) 133; *Mormoops megalophylla* (Ghost-faced Bat) 133; *Myotis* 129; *Myotis oxyotus* (Montane Myotis) 130; *Myotis pilosatibialis* (Hairy-legged Myotis) 130; *Noctilio leporinus* (Greater Bulldog Bat) 136; *Nyctinomops laticaudatus* (Free-tailed Bat) 129; Phyllostomidae with stripes 131; *Promops centralis* (Big Crested Mastiff Bat) 136; *Pteronotus fulvus* (Tawny Naked-backed Bat) 133; *Pteronotus mesoamericanus* (Mesoamerican Moustached Bat) 133; Reserva Biosphere de Calakmul 133; *Rhogeessa tumida* (Black-winged Little Yellow Bat) 133; *Vampyrum spectrum* (Spectral Bat) 134; volcano of bats 133